THE
CREATIVE
GARDENER

THE
CREATIVE
GARDENER

ADAM FROST

PHOTOGRAPHY BY JASON INGRAM

CONTENTS

PREFACE

Through the process of writing this book, I've come to realize something important: whatever you're making, be it big or small, complicated or easy, there's as much pleasure to be had from the process as from the end result.

The planning, researching, choosing the materials, learning new skills, and then the fun of putting it all together—that is where the real enjoyment is found. I also realized that I get a deep sense of calm and satisfaction when I use my hands to make something.

These days, it seems that we live in a world where it's all about the end goal, the destination. I want this book to celebrate the benefits of taking your time to enjoy the journey and making the connection to the simple pleasures of learning new skills, using your hands, and spending time productively. It doesn't matter if it takes you a weekend or three years to finish it, when you make something, you also create memories, and this is what gives your garden personality and soul.

Since way back when I was in my early twenties and worked with the late, great gardener and TV presenter Geoff Hamilton, I've been called a gardener, a landscaper, and a designer, but ultimately what my work is really about is being able to turn an idea into reality. My grandad, uncle, and old man always made things, and the projects in this book are just an evolution of that process. I like to think of them as part of an ongoing artisan tradition.

From knocking bits of timber together in my grandad's shed when I was a kid to the present day, I've always loved using my hands, but it's only recently dawned on me that when I make things today, it gives me a connection with my past. I've realized just how important making things is to my sense of self— it's how I fit into the world.

For anyone wanting to create something for their garden for the first time, I'd recommend flicking through the book and finding the thing that you're excited about making, rather than worrying too much about the skills. Even if you have a bit of a meltdown in the process, just stick with it, as every project here is achievable.

It's a hard thing for me to say out loud, but I'm really pleased with all these

projects. I'm aware that people who've not made anything before might look at some of them and think, "Well, I've never picked up a saw or a chisel, I can't do this, I can't do that." But none of the projects are very difficult, and I'd feel comfortable teaching my 14-year-old to do everything in this book. It's true that you might need to invest in a few tools or other pieces of kit but, more than anything, you need to invest in setting some time aside.

Perhaps rather than doing one of the projects by yourself, you might prefer to involve a friend or neighbor, or, even better, do it with your family. Or you could commission someone else to do it to your specifications and watch and learn as things take shape. If, on the other hand, you're more experienced

with a few skills under your belt, then I'm hoping these projects might provide a springboard for creating something entirely different.

My aim with this book is to provide a shopping list of materials and the selection of tools you need, then to give clear step-by-step explanations on how to make the projects. Hopefully, the book will give you not only the knowledge and the skills, but also the confidence to go ahead and actually make something. And not just make any old thing, but something that's really beautiful and useful.

CREATING ATMOSPHERE

USING YOUR TIME

RECYCLING AND UPCYCLING

FINDING INSPIRATION

INSPIRATION

> "Sometimes it helps to see things simply as forms rather than thinking of them as what they actually are."

INTRODUCTION

When I'm creating gardens, I always think about how they can bring about special moments and memories. Importantly, it's not only about the end result but also about the journey, and, for me, making things from scratch or recycling brings more atmosphere and meaning to your garden. It's just a matter of learning a few skills, getting a few tools together, and looking for inspiration.

The best way to find ideas is to pay more attention to what's all around you and then let it percolate. The ideas in this book have come from many different places.

The internet is, of course, a vast source of inspiration, but wandering around antique shops, museums, and art galleries, and looking at fashion and cars, can all be inspirational. The main thing is to take a look at what's come before. And when you do see something you like, think about why you like it. What is it that catches your eye—the quality, the color, the feel? What is it that's singing to you? How is it affecting your senses?

Often, when I spot something I like, the idea simply jumps from one thing to the next. So I might notice a tiny screw or a color combination and use it in my next project.

> *You can find inspiration in everything. If you can't, then you're not looking properly.*
>
> Paul Smith, designer

CREATING ATMOSPHERE

Anything and everything can contribute to creating atmosphere, from the quality of materials, colors, textures, and form, to whether you buy, handcraft, or recycle the elements. Where you place things and how light falls on them—be it sun, shade, or dappled light—are also important.

For me, atmosphere is created by lots and lots of tiny little elements, and all these things play their part. Often, it's about understanding the wider environment, how we react to stuff. Do we want something to be in full view or do we want it to be more tucked away so it creates an element of surprise? You can make a lovely coffee table, for instance, but if you put it in the wrong place, it won't sing in a way that it possibly could.

It's also worth thinking about how something is made and the emotional response it could evoke. Although people will have different emotional responses to things, the chances are they will have a more positive response to an object that's handmade and heartfelt rather than mass produced and shop bought. It's also the stories behind things that makes them meaningful, builds atmosphere, creates memories, and gives a garden a personality. Think about creating moments in the garden when something will catch your eye and you can think, "We had fun making that as a family," or "I made that." There's undoubtedly a huge sense of satisfaction to be had from all of this.

GO FOR CUSTOM

Also, I think that the moment you make something custom for your garden, you can feel that it's really yours and you'll get more joy from it. Choosing the color that you want, the proportions, taking the design in another direction, or putting a twist on it are all elements that add to an atmosphere that becomes unique.

All the projects we've included are really just a starting point. I'd really love to see pictures on social media where someone has picked up an idea and run with it, and made it bigger and better, changed the design somehow, or just used one of the skills, such as carving, to make an object completely different from anything else. Find me on Instagram @adamfrostdesign.

Creating atmosphere in the garden is really about considering all the elements that make the whole—colors, light, and position of objects all play a part.

USING YOUR TIME

Finding time to slow down and look around for ideas isn't always easy, but if you can, it's a great way to see the world in a new light. Observing the details in things and thinking about how you could incorporate certain elements into your garden is always inspiring.

Perhaps you've got more time on your hands because you're not commuting any more, the kids have left home, or you've retired. You might choose to spend that time in the garden or making things for it.

Hopefully, this book will show you the benefits of slowing down. In a way, it's the same process when you realize the importance of doing something yourself, rather than paying someone else to do it. We live in a world where it's so easy to buy, you don't even have to get out of bed. When you make things yourself, you begin to see the world through a slightly different lens.

WHY DO IT?

So why would you want to personalize your garden? There may be many reasons, but the main benefits come if you want your garden to be special or because you want to feel a connection to it, to feel it's truly yours.

Some of this process is quite old school, and might bring to mind things that our grandparents or parents used to do. Creating new memories is important too, and the more we can make our gardens ours and express our personality, the more memories we associate with them.

It doesn't matter whether you're experienced or not, it's about making a start and finding a way to relate to your garden. What's important is taking a step out of your comfort zone, and realizing that you can cut a piece of timber and make something out of it. Ultimately, the garden is for you, and I'd like this book to provide a starting point for you to connect to it.

I hope that this book will inspire you to take the time to slow down and think how you can create a personalized space.

RECYCLING AND UPCYCLING

Personalizing your garden is not just about making something from scratch—recycling and upcycling things that already have a history and patina are great ways to add more character to your garden.

I rarely throw anything away and love finding things in salvage yards that I can do up or repurpose. I think my love of salvaging stuff possibly comes from my childhood. The grandmother I used to call "scruffy nan" wouldn't throw anything away. She would use an old Belfast sink as a water feature, for instance. And my old man used to take me to Victorian dumps to dig for old bottles. The idea of finding something in someone else's rubbish, cleaning it up, and turning it into something useful and attractive really appealed to me. And I've still got a bit of a passion for Victorian bottles and inkwells even now!

REUSE, RECYCLE, REPURPOSE

To be honest, nowadays, upcycling couldn't be easier. You can do it from the comfort of your own bed by just trawling through the internet. There's loads of places selling or giving away stuff. Another reason I love recycling is that we live in a throwaway society and we really shouldn't. We should always ask ourselves if we can mend stuff, upcycle, or repurpose, and not just throw stuff away.

MAKING NEW OUT OF OLD

A number of the projects in this book use new materials, but many of them could be made from recycled materials. For instance, the compost bin on pages 222–27 could be made out of pallet wood or old scaffold boards. It's just about looking at what's available and using your imagination.

For me, making or restoring, recycling, and repurposing things is really rewarding and brings a sense of satisfaction that getting my wallet or credit card out simply can't touch. When I stand back and call the kids and Mrs. Frost to look at whatever I've made—or even better, if I can get them involved in making something—it really is one of the best feelings.

We found this bistro-style table and chairs online and gave them a new lease of life with a lick of paint (see pp142–45).

FINDING INSPIRATION

If you really start looking, you can find inspiration all around you. When something appeals to you, try to analyze why. Perhaps it's the texture, the material, or the color. Whatever it is, use this as a springboard and marry it with other ideas to come up with something that is unique and meaningful to you.

These simple log seats in a woodland setting (far left) make the perfect place to perch and take in the view. They were inspired by the simple beauty of logs or fallen tree trunks in a natural setting (left). Wood is such a lovely material in its own right, but using hand chisels to carve it is a great way of enhancing the grain or adding some subtle detail (below).

For me, watching birds feeding, drinking, and bathing (above) is one of the best things about improving the biodiversity of a garden. My aim with the carved birdbath (right) was to make it both beautiful and useful.

Narrow pathways (far left) encourage visitors to slow down and better appreciate a garden. Picking paths (left, above) are a great way of linking various areas of the garden, providing a quick route through planting, and are useful for doing maintenance too. The animal runs you can often spot weaving through hedges and fields (left) were the inspiration for the picking path I made.

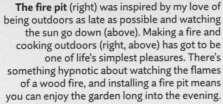

The fire pit (right) was inspired by my love of being outdoors as late as possible and watching the sun go down (above). Making a fire and cooking outdoors (right, above) has got to be one of life's simplest pleasures. There's something hypnotic about watching the flames of a wood fire, and installing a fire pit means you can enjoy the garden long into the evening.

Creating habitat such as this bee hotel (far left) is a good way of encouraging wildlife into your garden. By choosing pollen-rich plants, such as this poppy (left), we can do our bit to help reverse the recent decline in bee populations.

Inspired by the benches I saw in Japan, this wooden seat (right) seems low at first, but it's surprisingly comfortable and offers an interesting perspective on the garden. Nature provides an endless source of inspiration, such as these ripples in the sand (far right), which I had in mind when carving the sides of the bench.

My log store (far right) takes its inspiration from traditional Icelandic turf-roofed houses (below) and has a mini wildflower meadow on top of it, which protects the roof and also attracts beneficial insects. The Japanese technique of *shou sugi ban* (also known as *yakisugi*) involves charring the surface of wood (right) to waterproof and preserve it.

PLANTING AND
   PLANT SUPPORTS

WILDLIFE AND HABITATS

FURNITURE

WATER FEATURES

HARD LANDSCAPING AND
   DESIGN FLOURISHES

PROJECTS

> " *Having things set up in a dedicated work space, even if it's small, can be really helpful.* "

INTRODUCTION

Before you get started on the projects, it's worth equipping yourself with a basic tool set for soft and hard landscaping and for the workshop. I use chiseling to add detail to many of the projects in this book, so it's a good idea to invest in a small chiseling set too if you're planning on making those items.

I've listed the tools I find useful on the following pages, as well as the chisels I tend to use the most. You don't have to buy new tools if you don't already own them, as you'll probably find that you can borrow them from friends or neighbors, or perhaps from a local community gardening group. You can also sometimes find decent secondhand tools online or at yard sales and in thrift shops.

Even if you can only manage to work on projects for a few hours here and there, rather than bigger chunks of time over the weekend, I'd recommend setting up a work area for yourself as well as getting together a basic tool kit. That way, if you have a few spare hours, decide to take an extended lunch break, or just finish early, it's much easier to begin or continue one of the projects.

> " *A decent set of tools to work with gives you a head start on any building project, no matter its size.* "

BASIC TOOLS FOR SOFT LANDSCAPING

You'll need a good selection of cutting and digging tools to keep your garden looking its best. I tend to check all my tools over during the winter months, then give them all a good clean and a wipe with an oily rag.

CUTTING TOOLS

1. Loppers for cutting and pruning
2. Hoe for weeding and working the surface of the soil   **3. Lawn-edging tool** for keeping lawn edges neat
4. Garden shears for cutting hedging or long grass around trees   **5. Pruning saw** for cutting thick branches   **6. Knife** for cutting twine and taking cuttings
9. Japanese scissors for deadheading and cutting flowers   **11. Hand pruners** for pruning and cutting back

DIGGING TOOLS

7. Small planting spade for working small areas   **8. Transplanting trowel** useful for small plantings   **10. Hand fork** for loosening soil when weeding
12. Old Dutch-style planting trowel a heavyweight trowel for planting and loosening soil   **13. Spade** for digging, with treads at the top to protect boots
14. Border fork for lightly digging, lifting, turning, and aerating soil

OTHER TOOLS (not shown)

Garden rake for evening out soil and creating the right tilth, as well as for tamping down when laying turf

BASIC TOOLS FOR HARD LANDSCAPING

Certain tools are essential for building the solid structures in your garden. If you don't have some of these tools, you may be able to borrow them or find good-quality secondhand ones.

MISCELLANEOUS TOOLS

1. Bucket for measuring and carrying materials and water **2. Cordless drill** for drilling holes in wood and masonry (it's useful to have a range of drill bits and screwdriver bits)

BRICKLAYING TOOLS

3. Line blocks used with a string line to set up straight lines when laying bricks **12. Brick trowel** for laying bricks **14. Pointing trowel** for pointing mortar

MEASURING AND LEVELING TOOLS

4. Gauging trowel for applying mortar to tight areas **6. Level** for checking and setting levels over large areas **9. Carpenter's square** and builder's square for marking timber and checking corners **11. Boat level** for setting levels in smaller areas **16. Tape measure** **18. Line and pins** for marking out areas and setting levels

SPADES AND SHOVELS

5. Shovel for use with aggregates, sand, and cement **22. Small round-point spade** for digging in confined spaces

BRUSHES

7. Wire brush for cleaning stone and brickwork **21. Stiff hand brush** for general cleaning up

HAMMERS AND MALLETS

8. Rubber mallet for tamping down paving slabs and bricks **13. Claw hammer** has a claw for removing nails **15. Brick hammer** has a sharp chisel-shaped head for chipping off edges of bricks and stone **19. Hand sledge** for heavy-duty building work

SAWS AND CHISELS

10. Handsaw for cutting wood **17. Cold chisel** for chipping out small, precise details on concrete **20. Masonry chisel** for cutting bricks, concrete blocks, and paving

OTHER TOOLS (not shown)

Spray line for marking out lines
Mixing tray for mixing cement
Rake for moving fill dirt and gravel

SAFETY FIRST

Always wear **safety goggles, heavy-duty gloves,** and **steel toe-capped boots** when using these tools to protect yourself from injury, and make sure a **first aid kit** is available.

BASIC TOOLS FOR THE WORKSHOP

Having a kit of useful tools will help make the projects easier. Check what you need for the projects you're interested in, but many are useful for general work with wood and stone. Carving wood or stone is great fun, so get decent tools from the outset.

THE TOOLS

1. Metal ruler for accurate measuring
2. Wooden mallet for knocking pieces of wood together or driving in dowels
3. Pencil for marking shapes and measurements   **4. Sliding bevel** for setting and transferring angles
5. Hammer for striking things and knocking in nails   **6. Paint scraper** for scraping off paint and rust   **7. Hand plane** for shaving off thin layers of wood to make it smooth   **8. Stone-carving chisels** for refining the shape and adding details to stone   **9. Claw-bit holder with claw bit** for cutting into and creating texture in stone   **10. Metal dummy hammer** for striking stone chisels   **11. Gouge axe (aka adze)** for removing chunks of wood and carving, shaping, and hollowing wood accurately   **12. Spokeshave** for shaping and smoothing, particularly useful for curves or complex shapes in wood
13. Hand plane for smoothing surfaces, shaping wood, and reducing its thickness
14. Wooden drill bit for drilling cleanly through wood   **15. Countersink drill bit** to create a cone-shaped recess for screws or bolts to sit flush or below the surface of wood   **16. Carpenter's wood scribe** for scratching measurements on wood
17. Carpenter's square/set square for accurately setting a right angle.
18. G clamp for securing or holding work in place   **19. Carver's mallet** a cylindrical tool to use with a wood chisel or gouge   **20. Wood chisels** for shaping and carving wood

TYPES OF CHISELS

Using chisels to carve wood or stone takes a bit of practice, so it's best to experiment first and discover what you can achieve with the various shapes and sizes of chisel.

The moment you pick up a set of chisels, you'll most likely spend hours messing about with them and having fun. These images show what I've achieved with the various chisels. I use a carver's mallet for the wood chisels and a metal dummy hammer for chiselling stone. Remember that to get the best out of your chisels—or any cutting tools, for that matter—you'll need to keep them sharp by using the appropriate sharpening kits, such as a honing guide to set the correct angle and a sharpening stone.

This gouge chisel is the widest one I've got, and it's great for roughing out the main shape or making wide marks in wood. I used it on the toad house (pp112–15) and low bench (pp146–53).

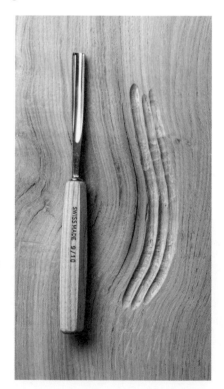

This 9/10 gouge chisel has a bit of a curve in it so it's one of the easiest to work with. You have to use the flatter chisels on an angle. I created the pattern on the low bench (pp146–53) with this one.

I used both of these chisels on the sparrow terrace (pp102–11). The small V chisel (far left) is great for scribing lines in wood. It came in handy for the sparrow terrace, giving me a line to work to in the wood. A flat chisel (left) is a good general use tool that can be used in lots of ways. I used it to widen and give a crisper finish to the line I made with the V chisel.

A gouge chisel comes in useful to work more detail into wood. I used this chisel for the sparrow terrace (pp102–11).

This curved gouge chisel has a broader blade than the 9/10 chisel (on facing page), so it creates a wider cut. I used it for adding finer detail to the toad house (p112–15), for making the front of the bee habitat (pp116–19), and for the pattern on the top of the simple log seat (pp138–41).

This stone-carving tool comprises a claw-bit holder and an interchangeable replaceable claw bit. You use it with a metal dummy hammer, angled either flat or at an angle, to create different effects. Here, I've used it in two directions, one at right angles to the other, to create a dimpled effect. I used this tool for the curved bird bath (pp96–101) and for the coffee table planter (pp132–37).

PROJECTS

PLANTING AND PLANT SUPPORTS

> *Growing plants in containers is a great way to experiment if you don't feel confident about planting design.*

INTRODUCTION

Learning about planting and planting design is really the journey of a lifetime. Here I've focused on growing in containers as it's a great way of experimenting, and is perfect for getting to know plants through the seasons and developing as a gardener. To give your garden a little style and personality, I've included what I hope are a few interesting and beautiful structures to use as plant supports.

When you grow plants in containers, you're not committing to anything, so it doesn't matter if you get it wrong. You can simply keep trying out new plants and combinations as a way of moving forward with your ideas. And it's a great approach to start collecting plants if you rent your property because you can take them with you.

Plant supports are a good way to personalize your garden, but I've seen so many unattractive ones. Most of the supports you can buy are all very similar, and it seems that they're one of the elements in our gardens that we don't always get right. The supports I've included here use natural materials, which instantly look more at home in the garden. You can customize them as you wish—you could paint the obelisk (see pp80–87), for example, to suit the palette in your garden.

> *The more we understand our plants, the more we can personalize our space, and growing plants in containers will help with that understanding.*

Peat-free commercial potting mix

Gravel

Composted bark

Topsoil

Sand

POTTING MIXES FOR CONTAINERS

When I'm putting in plants, the medium in which I choose to grow them is always driven by nature, by what the plant needs. So it's always worth doing some research to find out where the plant grows happily in nature and then try to emulate those conditions.

I have a store of the ingredients shown on the opposite page and mix them together as needed. It's not an exact science, and the best way to learn what plants need to thrive is by trial and error.

ALL-ROUNDER

For an all-purpose potting mix, I use 3 parts topsoil, 5 parts peat-free potting mix, and 2 parts sand.

WOODLAND PLANTS

For shade-loving woodland plants, I try to recreate the conditions of a woodland floor, where there's lots of leaf mold and bark. I use 2 parts commercial potting mix, 1 part composted bark, and 1 part topsoil. You could also add in half a part of gravel if you want to make the mix more free-draining.

FREE-DRAINING PLANTS

If the plants are more free-draining, I'd use a mix of 2 parts commercial potting mix, 1 part topsoil, and 1 part gravel.

LONG-TERM CONTAINERS

If I were planning for the containers to last a few months through the season, I'd usually add more topsoil to give the mix more body.

1 On a clean surface, measure out the parts of soil, commercial potting mix, and sand to make the potting mix.

2 Use your hands or a trowel to mix everything together thoroughly so it's nice and consistent.

3 Take a handful of the mix and scrunch it together to check the consistency. Adapt the mix to suit the plants you'll use.

PLANT CONTAINERS

Putting together container displays is a great way of trying out new plants before actually introducing them to the garden. I like using lots of single specimens in pots to create a pick and mix, so I can change and adapt things as the season progresses.

YOU WILL NEED

- homemade potting mix (3 parts topsoil, 5 parts commercial potting mix, 2 parts sand)
- terracotta pots/containers
- plants
- mulch, such as fine gravel or bark

Tools

- trowel
- small scoop
- soil sieve (for sifting out large lumps)
- watering can/hose

I use a soil-based, peat-free potting mix that I make myself as it gives weight and stability to the pot, and the plant invariably settles in better when I move it into the garden. Depending on a plant's needs, I might adapt the mix slightly. I'll add leaf mould to improve moisture retention, for instance, or sand to make it more free draining. When I'm choosing plants, I look at verticals, horizontals, and shapes and forms, as well as color, texture, movement, sound, and scent. Then I think about how much interest and atmosphere I can create. I recommend having a mix of different-shaped pots, but keep things simple by using only a few materials. It's a personal thing really, but I find terracotta works really well.

1 On a clean surface, measure out the parts to make the potting mix and combine together with your hands.

2 If there's one large hole in your container, you could put a few crocks, such as pieces of broken terracotta, or fine wire mesh over it to ensure the potting mix doesn't wash away.

3 Add your potting mix to the pot. Depending on the plants you decide to use, you may want to adjust the mix slightly. Cover the base of the container with potting mix so there's plenty for the roots to sit on and the plant will be positioned comfortably below the rim.

4 Take the plant out of its pot and gently tease out the roots if they're compacted. I've used a red grass called *Imperata cylindrica* 'Rubra' (classed as a weed in the US).

5 Then position the plant in the center of the container.

6 Top up the container with your potting mix, adding it around the edge of the plant and gently pressing it into place to remove any air pockets.

7 Top-dress by adding a layer of mulch such as fine gravel or bark to retain moisture, suppress unsightly weeds, and look smart. Then put the pot in its final position among the other plants in the display and water it thoroughly.

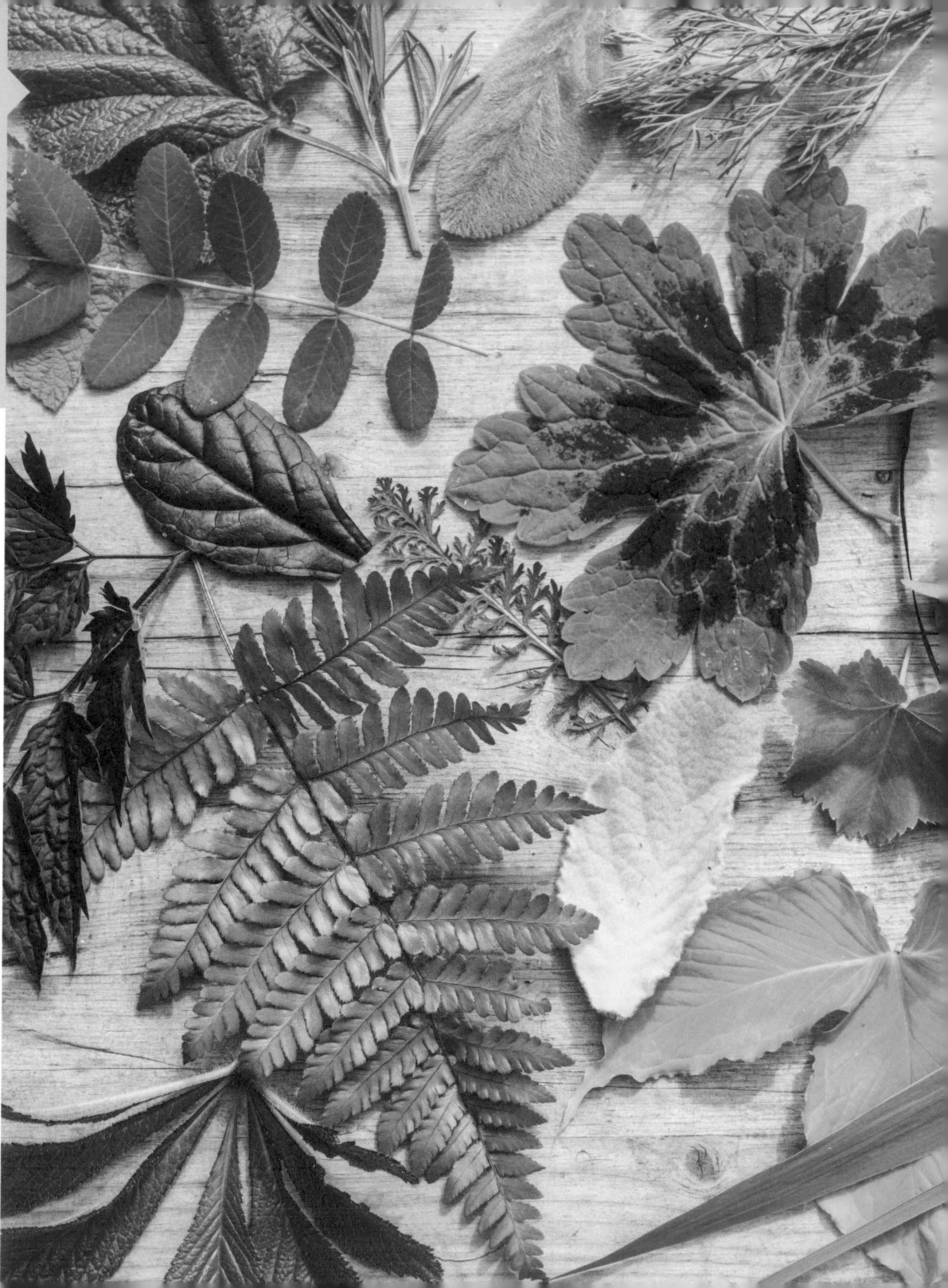

COLOR, TEXTURE, AND SHAPE

When it comes to planting, I'd like to encourage you to learn and experiment, and generally go about things with a bit more freedom in order to create what you want to create and not necessarily follow the crowd. We all see things differently and that is to be celebrated.

As well as choosing plants that catch your eye, it's good to take your time to look at how the groups of plants sit together, how the shapes formed by the horizontals and verticals play off each other, as well as the colors and textures. And one of the great things about using individual planting pots arranged as a group, as we've done in this book, is that you can experiment and change the display whenever you like, add things in or take them out. Whether you're a complete novice or a relatively experienced gardener, it's a great way of finding your own style and planting displays that put a smile on your face.

Paint chips (above) enable you to simply play with color. You will soon get a feel for what looks good. Building up your planting as layers of shapes, textures, and colors (left) is an easy method of understanding how plants can work with one another. There are lots of different leaf shapes (opposite) that you can add to your planting to provide even more layers of detail.

WOODLAND CONTAINER

This container relies predominantly on the texture, shapes, and forms of the foliage and how they play off against one another. The *Semiaquilegia* provides color among the fresh greens, but the container still holds interest even when the flowers have gone.

YOU WILL NEED

- galvanized container
- potting mix suitable for woodland plants: I have used 2 parts commercial potting mix, 1 part topsoil, 1 part composted bark, and half part gravel
- small scoop
- watering can/hose

Plants

- *Epimedium* 'Amber Queen'
- *Dryopteris stewartii*
- *Semiaquilegia ecalcarata*
- *Thalictrum* 'Purple Marble'
- *Polystichum setiferum*
- *Dryopteris sieboldii*

Dryopteris stewartii

Epimedium 'Amber Queen'

Semiaquilegia ecalcarata

Dryopteris sieboldii

Thalictrum 'Purple Marble'

Polystichum setiferum

1 First, I place the key plants from which to build the rest of the feature. Here I have used a fern, which provides great versatility and textural interest.

2 From there, the next thing I do is think about leaf shape and look for good contrast to my key plant. Here I am creating the backdrop to the rest of the planting. Work the potting mix around the roots of the plants as you go.

3 Once your backdrop has been formed, work from one side to the front and then repeat on the other side. Make sure you keep moving around to view your container from the front. Try to link plants across the container, which will help to create a rhythm and draw the eye.

4 When you have everything in place, make sure all the plants are well planted with their roots in the soil. Then, make your way around the plants, topping up the soil as you go. Gently firm the potting mix around the plants—the first watering will settle the soil around the roots.

Unlike the other collections of pots, the planting in this single container was designed purely to evoke a particular moment, a particular emotion.

LATE SPRING CONTAINERS

I chose the rich, bold colors of this collection of pots to work well against the background of old stonework and a wooden door. It's a scene that catches your eye and draws you toward it, rather than a place to pass through.

YOU WILL NEED

- containers
- peat-free potting mixes
- small scoop
- watering can/hose

Plants

- *Epimedium pubigerum*
- *Physocarpus opulifolius* 'Lady in Red'
- *Rheum palmatum* 'Atrosanguineum'
- *Dryopteris erythrosora*
- *Geum* 'Prinses Juliana'
- *Geranium phaeum* 'Lilacina'
- *Lamprocapnos spectabilis* 'Alba'
- *Epimedium x versicolor* 'Sulphureum'
- *Iris* Californian hybrid (Pacific Coast)
- *Mukdenia rossii* 'Karasuba'
- *Euphorbia cyparissias* 'Fens Ruby'
- *Libertia chilensis*

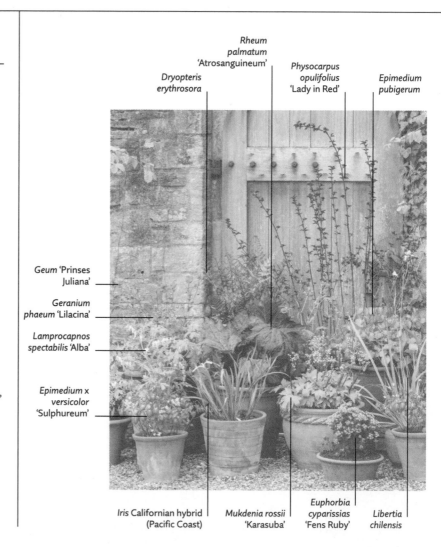

Rheum palmatum 'Atrosanguineum'

Dryopteris erythrosora

Physocarpus opulifolius 'Lady in Red'

Epimedium pubigerum

Geum 'Prinses Juliana'

Geranium phaeum 'Lilacina'

Lamprocapnos spectabilis 'Alba'

Epimedium x versicolor 'Sulphureum'

Iris Californian hybrid (Pacific Coast)

Mukdenia rossii 'Karasuba'

Euphorbia cyparissias 'Fens Ruby'

Libertia chilensis

When putting together this late spring display, I first positioned the dark-colored plants—the *Rheum* followed by the *Physocarpus* then the coppery *Dryopteris* fern—to create anchor points. Then I added other plants to soften the overall look and break up the darkness. It's useful to think about the hierarchy of the plants and the direction that the eye is drawn around the scene, and use this to create pace in the planting.

The dark leaves of the *Rheum* and the *Physocarpus* (left) bring out the bold orange color of the *Geum*. The white, heart-shaped flowers of *Lamprocapnos spectabilis* 'Alba' (known as bleeding heart or lady in the bath) (above) draw the eye and help to lift the darker plants.

Using terracotta containers (left and below) in varying sizes, styles, and finishes means that the eye focuses more on the plants and how the various colors, forms, and textures interact.

“ *This collection plays with the slightly richer, warmer colors that spring can provide rather than the more commonly used blues, whites, and emerald greens.* ”

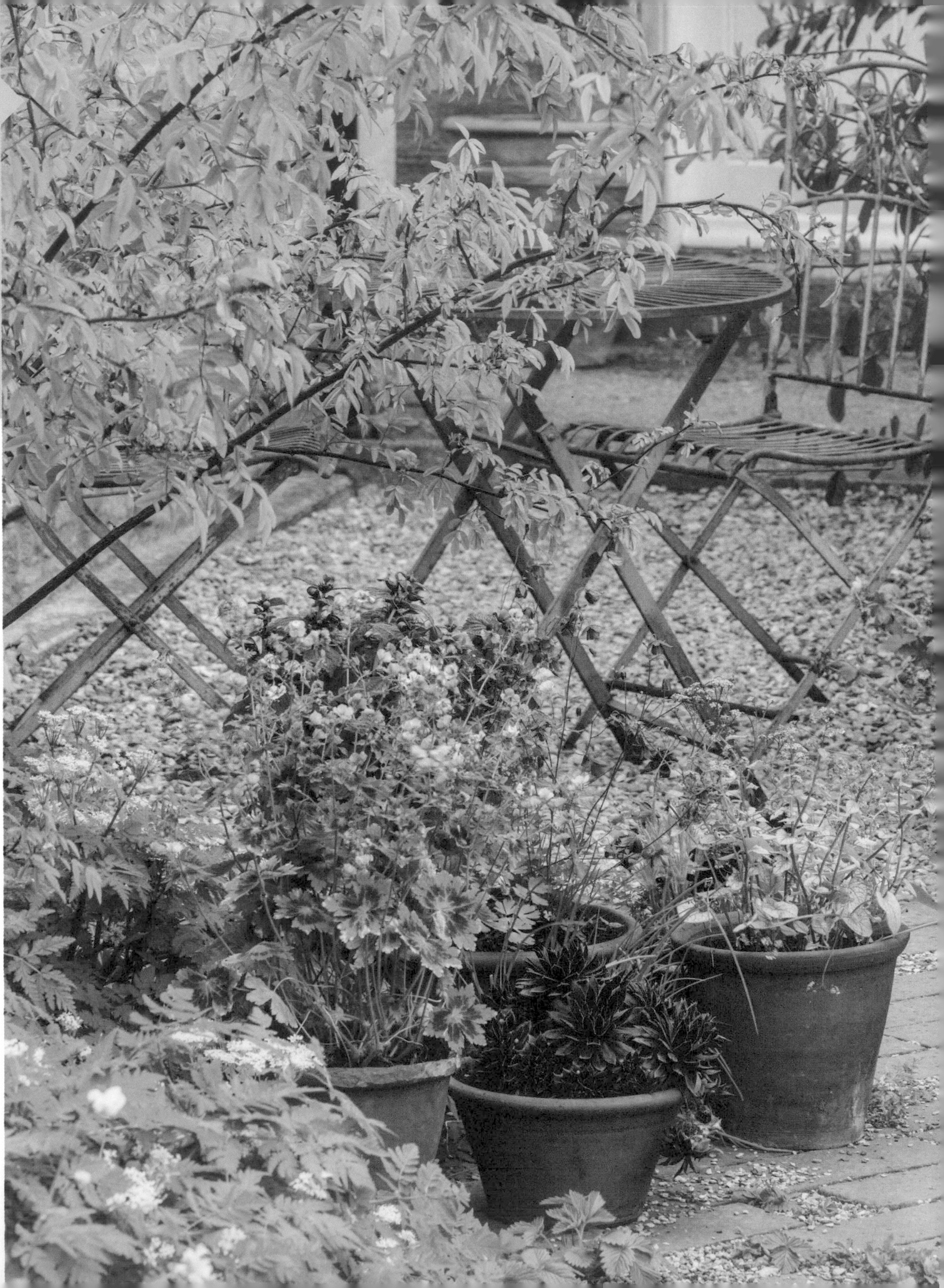

CONTAINERS FOR A SUNNY SPOT

I've laid out these pots so that they frame the space and encourage people to slow down. Sitting nearby and looking down on the plants, you'll find lots of interesting detail in terms of shape, color, and form.

YOU WILL NEED

- containers
- peat-free potting mixes
- small scoop
- watering can/hose

Plants

- *Lamium orvala*
- *Semiaquilegia ecalcarata*
- *Geranium phaeum* 'Misty Samobor'
- *Sempervivum* 'Pacific Purple Shadows'
- *Brunnera macrophylla* 'Jack Frost'
- *Primula pulverulenta*
- *Rubus arcticus*
- *Centaurea montana* 'Purple Heart'

SUIT THE SETTING

This arrangement was created to look good in spring, but none of the pots are too heavy, so you could easily pull some out and put new stuff in. When designing with plants, keep in mind that it's not just about using things that you love, but also considering what works together and fits comfortably into the wider setting. For instance, if I'd used exotics here, it would have jarred and looked really odd.

Centaurea montana 'Purple Heart'

Rubus arcticus

Primula pulverulenta

Brunnera macrophylla 'Jack Frost'

Lamium orvala

Semiaquilegia ecalcarata

Geranium phaeum 'Misty Samobor'

Sempervivum 'Pacific Purple Shadows'

The way the plants are laid out slows down visitors, and the plants' delicacy, detail, and rich colors add to the enjoyment of the journey through the space.

The rich, dark tones of the *Sempervivum* and the claret color of the *Primula* and *Semiaquilegia* help link the plants together, while the various leaf forms and texture add more interest.

" It's good to think about the setting as well as how you can make a connection to the next part of the garden. "

SUMMER CONTAINERS

This collection of perennials and grasses comes into its own in midsummer. The soft palette of colors and gentle movement of the grasses and *Diascia* give it a romantic feel that works well in the setting.

YOU WILL NEED

- containers
- peat-free potting mixes
- small scoop
- watering can/hose

Plants

- *Gillenia trifoliata*
- *Adenophora pereskiifolia* 'White Blaze'
- *Miscanthus sinensis* var. *condensatus* 'Cosmopolitan'
- *Digitalis* x *valinii* 'Firebird'
- *Thalictrum minus* 'Adiantifolium'
- *Penstemon barbatus* 'Roseus'
- *Astrantia major* 'Alba'
- *Geranium sanguineum* var. *striatum*
- *Linaria* x *dominii* 'Carnforth'
- *Salvia nemorosa* 'Bumblesky'
- *Trifolium ochroleucon*
- *Fragaria vesca*
- *Diascia* 'Hopleys'
- *Valeriana sambucifolia*
- *Stipa gigantea*
- *Calamagrostis* x *acutiflora* 'Karl Foerster'
- *Penstemon* 'Hidcote Pink'
- *Saxifraga* 'Kinki Purple' (*stolonifera*)

Miscanthus sinensis var. condensatus 'Cosmopolitan'

Adenophora pereskiifolia 'White Blaze'

Gillenia trifoliata

Digitalis x valinii 'Firebird'

Thalictrum minus 'Adiantifolium'

Penstemon barbatus 'Roseus'

Astrantia major 'Alba'

Geranium sanguineum var. striatum

Linaria x dominii 'Carnforth'

Salvia nemorosa 'Bumblesky'

Trifolium ochroleucon

Fragaria vesca

I've positioned this mix of terracotta and galvanized metal pots near a gateway, and chosen colors that sit comfortably with the surrounding gravel and stonework as well as with the plants glimpsed on the far side of the gate. A few pots near a gateway or a pause point in the garden add layers to your movement around the space. Groups like this can also be used as focal points in their own right or split and positioned either side of a pathway.

iascia 'Hopleys'

Valeriana sambucifolia

Stipa gigantea

Calamagrostis x acutiflora 'Karl Foerster'

Penstemon 'Hidcote Pink'

Saxifraga 'Kinki Purple' (stolonifera)

Geranium sanguineum var. striatum

Salvia nemorosa 'Bumblesky'

" *One of the great things about creating a display with containers is that, as plants fade, you can lift them out and add other things in to keep everything looking fresh.* "

Think about where your display goes, as often it is the material of the backdrop that brings the plants alive. Here, the wall and planting help create a country-cottage feel (above). Aim for a combination of shapes and forms, which offers good contrast, while repeating color provides rhythm (left).

When choosing plants, work out how the different shapes and forms best sit together. For example, try solid, rounded shapes next to light, transparent things, or a mass of close-growing flowers beside tall, straplike leaves. Whatever you choose, always try to get a balance between the vertical and horizontal elements. Also take into account how the colors of the flowers and leaves work together, and how these relationships will change over the season. Think about the details and smaller relationships within the bigger picture.

The pale flowers of *Trifolium ochroleucon*, *Astrantia major* 'Alba,' *Gillenia trifoliata*, and *Valeriana sambucifolia* add a lightness to the display, while the terracotta tones of the *Digitalis* x *valinii* 'Firebird' and pink of the *Penstemon barbatus* 'Roseus' bring warmth.

LATE SUMMER CONTAINERS

I've put this late season display next to my office door, so I get pure joy out of the plants every single day. For late summer interest, I'm not just looking at flowers, but also foliage, berries, and seedheads.

YOU WILL NEED

- containers
- peat-free potting mixes
- small scoop
- watering can/hose

Plants

- *Imperata cylindrica* 'Rubra'
- *Gaura lindheimeri* 'Rose Fan'
- *Miscanthus sinensis* 'China'
- *Kniphofia* 'Fiery Fred'
- *Actaea simplex* (Atropurpurea Group) 'Brunette'
- *Verbena bonariensis*
- *Viburnum plicatum* f. *tomentosum* 'Mariesii'
- *Dryopteris filix-mas*
- *Pachyphytum glutinicaule*
- *Fuchsia* 'Hawkshead'
- *Echeveria agavoides* 'Ebony'
- *Begonia sutherlandii*
- *Lychnis flos-cuculi* 'White Robin'
- *Aeonium balsamiferum*

Miscanthus sinensis 'China'

Imperata cylindrica 'Rubra'

Kniphofia 'Fiery Fred'

Gaura lindheimeri 'Rose Fan'

Actaea simplex (Atropurpurea Group) 'Brunette'

Verbena bonariensis

Viburnum plicatum f. *tomentosum* 'Mariesii'

Dryopteris filix-mas

Pachyphytum glutinicaule

Fuchsia 'Hawkshead'

Echeveria agavoides 'Ebony'

Begonia sutherlandii

Lychnis flos-cuculi 'White Robin'

Aeonium balsamiferum

There are about 25 plants in this display, including grasses, succulents, shrubs, ferns, and perennials. A couple of the stars are the red grass *Imperata*, which is like fire coming out of a pot, and the *Miscanthus sinensis* 'China.' This reaches about 4ft (1.2m) tall and works really well with the *Kniphofia*, *Gaura*, and *Verbena bonariensis*. These grasses look great, and they also look beautiful when they've died back. Remember, it's important to consider how your plants will look when they die back at the end of the season.

Rudbeckia triloba

Agastache 'Blackadder'

Eupatorium capillifolium

Symphyotrichum novi-belgii x turbinellum

Miscanthus sinensis 'Nippon'

Chloranthus japonicus

Imperata cylindrica 'Rubra'

Aeonium 'Velour'

Euphorbia mellifera

Fuchsia microphylla

Echeveria 'Apus'

Fuchsia 'Hawkshead'

Echeveria 'Mexican Giant'

Kalanchoe thyrsiflora 'Bronze Sculpture'

Plants

- *Rudbeckia triloba*
- *Agastache 'Blackadder'*
- *Eupatorium capillifolium*
- *Symphyotrichum novi-belgii x turbinellum*
- *Miscanthus sinensis 'Nippon'*
- *Chloranthus japonicus*
- *Aeonium 'Velour'*
- *Kalanchoe thyrsiflora 'Bronze Sculpture'*
- *Echeveria 'Mexican Giant'*
- *Echeveria 'Apus'*
- *Fuchsia microphylla*
- *Euphorbia mellifera*

You'll find that some plants are beautiful throughout the year, whereas others will fade and need replacing. What is the star at one moment may become an interesting backdrop later on. Be mindful to enjoy plants not only at their peak but also later on for what else they bring to the party, be that seedheads, berries, or foliage. The idea is that the display is as interesting as possible for as long as possible.

The thick leathery leaves of *Aeonium balsamiferum* make a good contrast to the fine foliage of the *Imperata cylindrica* 'Rubra' (below). Repeating the shapes, colors, and forms of plants, such as the upright stems of grasses, the fleshy leaves of succulents, and the yellows of the *Kniphofia* 'Fiery Fred' and *Rudbeckia triloba* (right), creates a sense of cohesion to the display.

> *One of the great things about growing plants in containers is that you can learn how they perform before making the commitment of planting them in the garden.*

Rudbeckia triloba and *Agastache* 'Blackadder' (left)—both of which bloom from midsummer to late fall—are prime examples of plants that keep displays looking interesting for an exceptionally long time. As well as considering how the display looks when facing it (above), think about how the plants work together from different angles, particularly when you move past them.

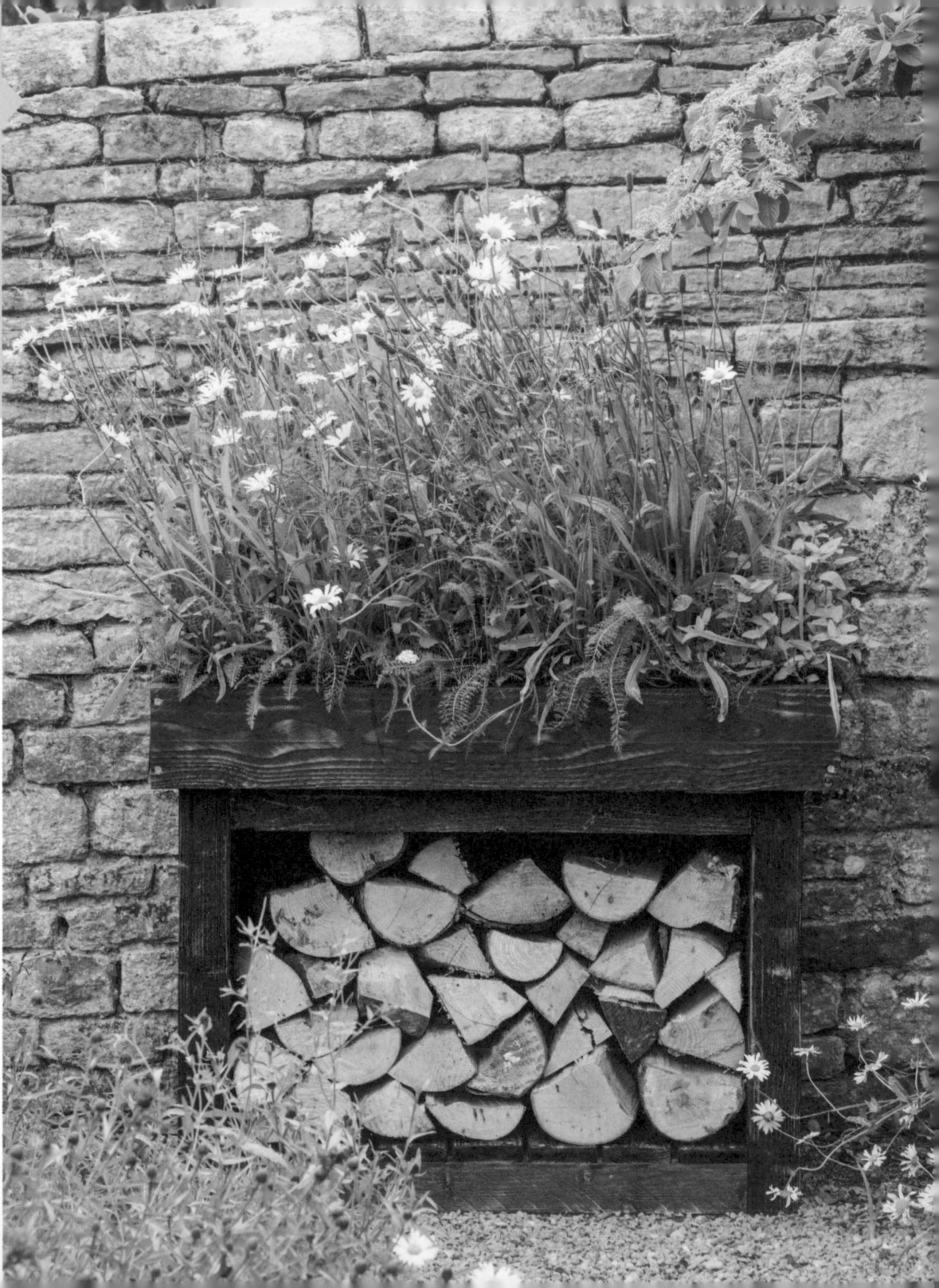

GREEN ROOF LOG STORE

This rectangular log store has a simple framework
covered with sawn softwood boards attached
horizontally all the way around, and I've experimented
with the Japanese technique of charring the wood,
called *shou sugi ban* (also known as *yakisugi*), to
blacken and protect the surface.

YOU WILL NEED

**See p72 for timber requirements
and measurements**

- 1½in, 2in, 2¾in, and 4in (40mm,
  50mm, 70mm, and 100mm) screws
- 59 x 47in (1,500 x 1,200mm) rubber
  or plastic pond liner
- 39 x 24in (1,000 x 600mm) roof drain
  (filter fleece layer, drainage layer, and
  protective layer)
- green roof substrate (lightweight
  aggregate and topsoil) laid 3–4in
  (75–100mm) deep (available from
  specialist suppliers and online)
- wildflower sod

Tools

- set square
- metal ruler
- saw
- pencil
- tape measure
- cordless drill/screwdriver
- ¼in (6mm) drill bits
- blowtorch
- sharp knife and scissors
- board (to use as a straightedge)
- piece of timber (for tamping)
- trowel
- watering can or hose

We all have to store stuff in the garden, and this is not only a more attractive way of doing so but is good for wildlife too. I've made this log store about 39in (1m) long and 24in (600mm) deep. The overall height of the front is 31in (780mm) and the back is 32½in (830mm), so the roof slopes forward slightly to allow rainwater to run off. The top forms a deep tray into which I've put a lightweight substrate and a strip of wildflower sod. Wildflower sod is a great way of creating an integral mini-meadow, and not only promotes biodiversity but also provides a good range of color and a long season of interest (usually from April to September).

CONSIDERATIONS

- You can change the proportions to make the log store longer and wider, but the techniques and skills used could also lend themselves to creating a tool store, trash can store, or bike shed, for example.

- You could also build a log store with a green roof on top as a way of dividing the garden into different areas or "rooms."

- If you're inspired to put a green roof on an existing structure, such as a shed, then bear in mind that the structure may need strengthening to take the extra weight.

- If you sit the structure on gravel or paving, then a base isn't necessary, but you could make one out of timber if you prefer.

EXPERT *INSIGHT*

- You can buy the components for the roof layers as a kit.

- It's good to cut all the frame timber to make as a "kit" rather than piece by piece. Then lay out the framework on a level surface.

- There are lots of options for planting, including sempervivums, sedums, grass, and chamomile. Here, I've used wildflower sod, which is a mix of wildflowers, herbs, and perennials.

- Use a substrate to suit your plants. For instance, the substrate needs to be very poor for wildflowers.

TIMBER FOR THE LOG STORE

Use the pattern below as a guide to measure, mark, and cut your cypress timber. The frame timber is cut slightly longer to start with and later cut to size. Only cut the front and back roof fascias after checking the lengths in step 13.

CROSS-SECTION OF THE WILDFLOWER SOD

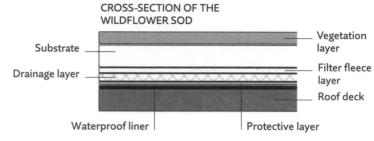

Substrate

Drainage layer

Waterproof liner

Vegetation layer

Filter fleece layer

Roof deck

Protective layer

INTERNAL FRAME
2 x 2in (50 x 50mm) thick

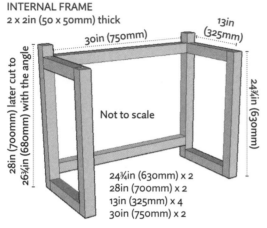

30in (750mm)

13in (325mm)

28in (700mm) later cut to 26¾in (680mm) with the angle

24¾in (630mm)

Not to scale

24¾in (630mm) x 2
28in (700mm) x 2
13in (325mm) x 4
30in (750mm) x 2

SIDE OF INTERNAL FRAME WITH FRONT UPRIGHT

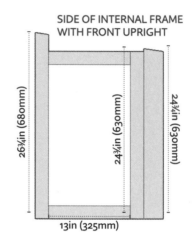

26¾in (680mm)

24¾in (630mm)

24¾in (630mm)

13in (325mm)

FRONT UPRIGHTS x 2
3in (75mm) thick

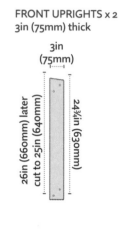

3in (75mm)

26in (660mm) later cut to 25in (640mm)

24¾in (630mm)

FRONT BEAMS x 2
3in (75mm) thick

30in (750mm)

3in (75mm)

BACK PANELS x 7
¾in (20mm) thick

34in (860mm)

4in (100mm)

FLOOR PANELS x 7
¾in (20mm) thick

19in (480mm)

3¾in (98mm)

SIDE SLATS x 14
¾in (20mm) thick

16¾in (425mm)

4in (100mm)

ROOF PANELS x 7
¾in (20mm) thick

22in (560mm)

4in (100mm)

SIDE ROOF FASCIAS x 2
¾in (20mm) thick

22in (560mm)

6in (150mm)

ROOF FASCIAS x 2
¾in (20mm) thick

37in (940mm)

6in (150mm)

FRONT FRAME
3 x 3in (75 x 75mm) thick

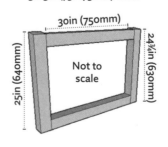

30in (750mm)

25in (640mm)

24¾in (630mm)

Not to scale

INTERNAL FRAME WITH FRONT FRAME ATTACHED

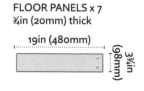

Not to scale

1 Cut the wood for the internal frame as shown on the opposite page. To create the angle for the roof, lay out one side of the internal frame, as shown here and in the diagram opposite. Then place one of the front uprights next to it at the front. Measure and mark the outer edge of the front upright at 24¾in (630mm) and the back of the internal frame at 26¾in (680mm). Use a straightedge and a pencil to mark the line of your roof angle across the timber.

3 Once you have cut your angles for the roof line, you can start to put your internal frame together using 2¾in (70mm) screws. Start by fixing the left-hand side, making sure the angled cuts at the front are at the top. Repeat the process for the right side.

2 Then cut across the pencil line with a saw. Repeat this process for the other side of the internal frame and front upright.

4 Next, use the 3 x 3in (75 x 75mm) thick timber to make your front frame. Fix the two front uprights and the front beams together, using 2¾in (70mm) screws.

5 Lay your front frame face down on the floor. Now you can start to fix your internal frame to the inside of the front frame with 2¾in (70mm) screws. This allows room on the outside to fix your paneling.

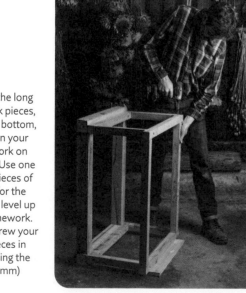

6 Fix the long back pieces, top and bottom, then turn your framework on its end. Use one of the pieces of timber for the sides to level up the framework. Then screw your back pieces in place using the 2¾in (70mm) screws.

7 Measure and cut the back panels, and then put them in position, making sure the ends are flush with your frame. Use 1½in (40mm) screws to fix each panel in two places to help stop the timber twisting. When you attach the side slats, they will hide the cut edges of the back panels.

8 Next, starting at the bottom, fix the side slats in place, using two 1½in (40mm) screws at either end of each slat. For the topmost slat, you will need to cut an angle that runs from the back (which is higher) to the front (which is lower). To do this, offer up the slat into position, then make a pencil mark at the front and back where it meets the uprights.

9 Using your straightedge, draw a line between the two pencil marks. Cut the timber along the pencil line and fix in place with 1½in (40mm) screws.

11 Next, you'll need to create the roof box for the planting on top. Offer up a roof panel at the side of the roof. Make sure you have an overhang of 2in (50mm) at each side, 2¾in (70mm) at the front, and 1½in (40mm) at the back. Fix one roof panel at each end of the roof, using two 1½in (40mm) screws at either end of each panel.

10 Now fix the floor panels into position, using two 1½in (40mm) screws at each end. Make sure they are square to the front and leave a small gap between each panel so that air can circulate.

> *The principle for building this structure can also be used for making other things that have different proportions or are bigger in scale, such as a trash can store or bike shed.*

12 Next, using 1⅛in (40mm) screws, screw the two side roof fascias into place, making sure they are flush with the base of the roof.

14 Once your front fascia is in place, you can fix in place the rest of the roof panels. Work from both sides toward the center, fixing with 1⅛in (40mm) screws at either end. You may find you have to mark and cut your final piece to size once the others are in place as it may not be a full piece. When the roof base is complete, fix the back fascia in the same way that you fixed the front.

13 When the side fascias of the roof box are fixed in place, measure from the outside edge of one side to the outside edge of the other to check the length for your front and back fascias of the roof box. Once cut, fix the front roof fascia in place by screwing it into the cut edges of the side fascias with 1⅛in (40mm) screws. This will give a tidy finish as it covers the cut edges of the sides of the box.

These log stores are really handy placed next to a grill, fire pit, or pizza oven, where you can get easy access to your wood store.

16 Place the log store in its final position before finishing your green roof as it does get a little heavy when filled. Line the roof with your pond liner, making sure it is folded neatly and overhangs each side.

17 The green roof is built up in layers. Once the waterproof liner is in place, you can cut and put the roof drain in position. It should be made up of a protective layer, a drainage layer, and a filter fleece layer (refer to the diagram on p72).

15 Using the blowtorch, start to scorch your timber. This finish can be used to create different effects. Make sure you are in the open with good air circulation.

18 Once you've folded all the layers neatly in place, fill with green roof substrate and use a piece of timber to level it. Make sure the roof is filled to the top.

19 Trim along the top edge of your outward-facing boards, removing any excess fleece and liner with a sharp knife. Don't worry if it's not spot on as you can always tidy it up when your roof is finished.

20 Measure up your final roof surface area and then cut your sod to this size. I use a board to help me get a straight edge.

21 Lift your sod into place, making sure it all sits on your substrate, like placing a large mat.

22 Tamp the sod into place using a piece of timber. Remember that if the weather is dry, it's a good idea to water your roof until the roots grow firmly into the substrate.

OBELISK

This obelisk makes a great climbing frame for plants or a feature—either stand-alone or in a series—to draw the eye or create a sense of rhythm. I added a finial at the top, which makes the whole thing look more dynamic and neatly hides the fixings.

YOU WILL NEED

See p82 for timber requirements and measurements

- 2 and 2¾in (50 and 70mm) brass screws

Tools

- workbench or flat table
- saw
- set square or carpenter's square
- tape measure
- pencil
- sliding bevel
- cordless drill/screwdriver
- steel ruler
- ⅛in (4mm) drill bit
- stepladder
- blowtorch

This obelisk is straightforward to build as you cut all the parts first and put them together like a kit. The most complicated section is working out the angles for the finial, but I've provided a diagram that will help you to lay things out (see p82). I've finished the wood by lightly burning the surface to create a charred finish, which works well with most color schemes and looks great covered in plants. Alternatively, you could coat it with linseed paint in a color that suits your garden.

CONSIDERATIONS

- Obelisks are great to use as a focal point. They give you instant height, and if you put two or three of them in a vegetable garden or herbaceous border, the repetition helps draw the eye and pull you through the space.

- It's an ideal structure for growing roses, sweet peas, or clematis on. Just make sure that whatever you grow isn't going to be too vigorous and swamp it.

EXPERT *INSIGHT*

- I will often use a plane to take the edges off the sides of the wood to give it a softer, more finished look.

- You could add a couple of pegs to the finial at the top to hang fat balls for birds to feed on.

- Linseed paint is great for outdoor structures as it helps maintain the wood and is highly durable.

- If you decide to paint the obelisk, choose a color that ties in well with other materials and the aesthetic of your garden.

TIMBER FOR THE OBELISK

Cut all the pieces of wood to the measurements given below so that you can assemble the obelisk much as you would a kit:

4 lengths 2 x 2in (50 x 50mm) cypress timber cut to 71in (1,800mm) for uprights

16 lengths ⅝ x 1¼in (15 x 30mm) cypress timber for crossbars. These will be cut to their final length in steps 5 and 11:

    4 at 13in (330mm); final length 9in (230mm)
    4 at 17in (420mm); final length 12¾in (320mm)
    4 at 20in (510mm); final length 16½in (410mm)
    4 at 24in (600mm); final length 20in (500mm)

¾in (20mm) thick cypress timber cut to 6¾ x 6¾in (170 x 170mm) for large top plate

¾in (20mm) thick cypress timber cut to 5½ x 5½in (140 x 140mm) for small top plate

4¾ x 4 x 4in (120 x 100 x 100mm) cypress timber for finial

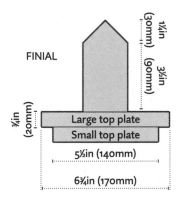

FINIAL

1¼in (30mm)

3½in (90mm)

¾in (20mm)

Large top plate
Small top plate

5½in (140mm)

6¾in (170mm)

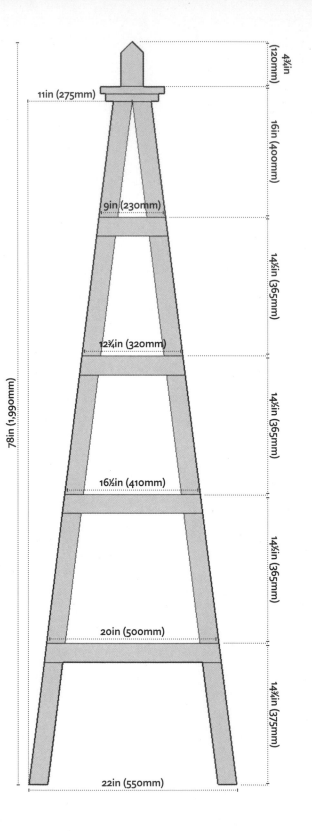

4¾in (120mm)

11in (275mm)

16in (400mm)

9in (230mm)

14⅜in (365mm)

12¾in (320mm)

14⅜in (365mm)

16½in (410mm)

14⅜in (365mm)

78in (1,990mm)

20in (500mm)

14¾in (375mm)

22in (550mm)

3 To establish the angle at the top of the uprights, the distance from the wood clamped on the side of the workbench to the top center of the uprights where they meet should be 11in (275mm). Once you have your uprights set out, it makes sense to draw around them with a pencil. This will help to make sure nothing moves.

1 Cut your four uprights to 71in (1,800mm) long. Lay them flat and use a carpenter's square and pencil to mark the position of the crossbars on all the uprights (refer to the diagram opposite). You may find it easier to mark two at a time, side by side. Next, cut the crossbars to length. Initially, these will be slightly longer than the finished length and are cut down in steps 5 and 11.

2 First clamp a length of timber to the edge of your workbench. You'll use this as a fixed point when establishing the angle for the uprights. Lay two of the uprights on the workbench with the tops butted together. To set the angle, position the bottom of the uprights 22in (550mm) apart, measuring from the outside edges.

4 Once you have the angles set, start to fix your crossbars according to the marks you made earlier. Take your time and make sure nothing moves. Predrill holes, one on each side of the crossbar, making sure the fixing goes through the center of the upright, then screw firmly into place using the 2in (50mm) screws.

5 Once all the crossbars are secure, turn the uprights over and use a saw to cut the wood overhanging at the sides. You can use the upright to guide the saw. The edges of the crossbars should sit flush with the uprights.

7 With the angle you set in step 6, place the sliding bevel 2in (50mm) from the top of the uprights and make a pencil mark across the two pieces of wood. This will create a flat surface onto which you can fix the top plates and finial.

8 Cut the top of the uprights along your pencil mark to create a flat surface.

6 Use a sliding bevel to get the angle between the crossbar and the upright posts. You will use this angle to mark the top of the uprights where the finial will go.

9 Repeat steps 2 to 8 for the other two uprights and crossbars. Turn the two frames on their side with the top ends butted together. Set the distance between the bottom of the uprights 22in (550mm) apart as you did in step 2. Then, using the pencil marks you made earlier as your guide, put the crossbars in place and clamp them to the uprights to hold the framework together. Double-check the measurements and angles match on all four sides of the obelisk and adjust if need be.

10 Once you've checked that the angles are correct and the crossbars are aligned, pilot drill the holes to fix them with screws (as you did in step 4). First, fix the top and bottom bars in place as these will hold the frame together. Then attach the remaining crossbars. Once complete, turn the frame over and repeat the process on the other side.

11 Using the side of the upright as a guide, repeat step 5 to saw off the overhanging wood so the sides of the crossbars sit flush with the uprights.

12 Take your 6¾ x 6¾in (170 x 170mm) square of timber for the top plate and mark the diagonals to find the center point. Repeat with the 5½ x 5½in (140 x 140mm) square.

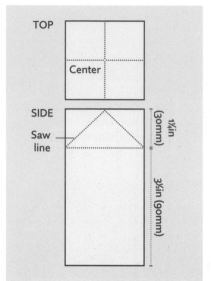

15 Take your 4¾ x 4 x 4in (120 x 100 x 100mm) piece of timber for the finial. Find the center of the block by marking two lines equidistant from the sides on the top. On one side of the block, mark a horizontal line 1¼in (30mm) down from the top. Next, mark two diagonal lines from the top center to either end of the horizontal line. These are your saw lines.

13 Position the 5½ x 5½in (140 x 140mm) top plate on top of the uprights, making sure it is central and square.

14 Clamp the top of the uprights to keep them butted up tightly together and then fix the small top plate to the uprights with four screws.

16 Mark two opposite sides as in step 15. Then clamp the block to your workbench so that you can cut along the saw lines vertically. You can do this with a handsaw but it's easier to achieve with a chop saw to get the angles correct. First, cut the two opposite sides. Unclamp the block and measure and mark the saw lines for the other two sides by drawing diagonal lines from the top center to the outer edges on both sides.

17 Reposition the finial in the clamp again so that you can make the cut vertically. Saw off the other two triangular shapes to form a pyramid shape at the top of the finial.

19 Position your finial and large top plate so they are square on the lower plate. Then pilot drill on two sides and fix with 2¾in (70mm) screws.

18 Mark the center point in the bottom of the pyramid finial and predrill a hole. Also predrill the center of the 6¾ x 6¾in (170 x 170mm) top plate. Clamp the finial to your work bench with its base pointing up, line up the holes, and screw the two pieces of wood together. The finial should sit in the center of the base.

20 Make sure you have plenty of space around you and ideally work outdoors. Use your blowtorch to char the surface of the wood. You can make it as light or as dark as you like. Work slowly and keep assessing it to make sure you're happy with the results.

HAZEL LADDER

These hazel stick ladders were inspired by the tall
ladders that you find in old libraries and bookshops.
They are a great way of training roses to climb up
into trees, and you can lean two frames against
each other to create a stand-alone feature for
growing beans or clematis.

YOU WILL NEED

- hazel poles: 2 uprights about 6–6½ft (1,800–2,000mm) long x 1in (25mm) wide, plus extra for rungs ½–⅝in (10–13mm) thick
- garden twine

Tools

- tape measure
- pencil
- pruners, saw, or loppers (for cutting timber)
- workbench
- chisel, about ¾in (20mm), or a small axe or knife
- mallet
- knife or scissors
- spade

I generally use hazel poles because they tend to be uniformly straight and are fairly strong. They last around four to five years in the garden and look good as they age. I use ordinary garden twine to tie the rungs to the uprights, but you could use any type of strong twine or cord and make a feature of it. I taper my ladders slightly—I usually make the bottom of the ladder 18–24in (450–600mm) wide and the top, where the last rung is, around 12in (300mm) wide, with the rungs spaced about 12in (300mm) apart.

CONSIDERATIONS

- Hazel sticks work well for this project, but you could also use metal rods, bamboo canes, or roofing battens, as long as the pieces are straight and long enough, and you can cut shorter lengths for the rungs.

- These ladders work best in ageing trees that are quite open. Plum trees or apple trees work well because they can be pruned to an open "goblet" shape.

- I tend to use rambler roses because, once established, they will tolerate competition at the roots and don't need much pruning to keep them looking good. Before planting your rose, make sure the area is clear of weeds. To give it a head start, mulch around the base to retain moisture and suppress weeds.

EXPERT *INSIGHTS*

- If you're using this ladder to train a rose into a tree, position it about 20in (500mm) away from the trunk. When your rose needs pruning, just cut stems at the bottom of the ladder and pull them out from the tree.

- One of my favourite rambling roses is 'Rambling Rector.' It's really vigorous and I particularly like its name!

- Lean one of the uprights against the tree to see what it looks like in place. If you make it slightly too big, you can always adjust it if necessary to make it smaller.

1 Measure and mark the height you want your ladder's uprights, making sure the dimensions suit its intended position. Then cut two poles to length using sharp pruners, a small saw, or loppers, depending on the thickness of the wood.

2 Hold the poles for the uprights firm by clamping them onto a workbench or have someone hold them for you on a flat surface. Then use a chisel and mallet to slice away the timber to create a point at the base of each one. This will help you to push the uprights into the ground.

3 Lay the uprights on a bench or on the ground and position them to the desired width and angle of your ladder. Make sure the tops and bases are flush with each other.

4 Cut a length of twine to bind the tops of the uprights together. Tie a loop slip knot and slide it over one upright. Then weave the twine in a figure eight pattern where the uprights cross. Once it feels tight and secure, tie it off with a simple double knot.

5 Cut the lowest rung first. Aim for an overhang of 1½–2in (40–50mm) either side of the uprights. Work out the distance you want between the other rungs and cut those to length too. My first rung is about 1ft (300mm) from the bottom of the ladder.

6 Position the lowest rung on the uprights and fix it in place using garden twine. Use a figure eight to bind the pieces together securely. Repeat the process with the remaining rungs, making sure they are parallel with each other.

7 Finally, position the finished ladder, leaning it against your tree at a slight angle. Push the points of the uprights into the ground to help keep it in position.

8 I planted my rambling rose at the base of the ladder to train it into the tree. You can do this before you position the ladder if you prefer.

" This simple ladder is a great way of adding personality to your garden and only takes 30 or 40 minutes to build. "

WILDLIFE AND HABITATS

> *We tend to want tidy gardens, yet letting things be a bit more scruffy is generally better for wildlife.*

INTRODUCTION

With loss of natural habitats and declining insect, bird, and animal populations, gardening with wildlife in mind is more important now than ever. The more we can provide habitat, the better—not only because it's good for the environment and boosts biodiversity, but also because it helps us slow down, be present in the moment, and appreciate the natural world.

I get a lot of pleasure from building these habitats, but the best moment is when I see them being used. In this section, you'll find out how to make a stone birdbath, terraced housing for sparrows, a toad house, a bee habitat, and a sculptural seat with lots of little niches for all sorts of insects. All the projects were inspired by me being drawn to certain creatures—whether that's through reading about declines in populations or just because I enjoy seeing them. Generally, gardeners tend to be quite neat and consider gardening for wildlife as being a bit scruffy, so this is my take on getting wildlife into a garden in a more acceptable way—making habitats that are both beautiful and useful. It's amazing to see how they can change the dynamics of a space, both for wildlife and for you.

> " *I created these features as habitats, but they also bring elements of design interest into the garden.* "

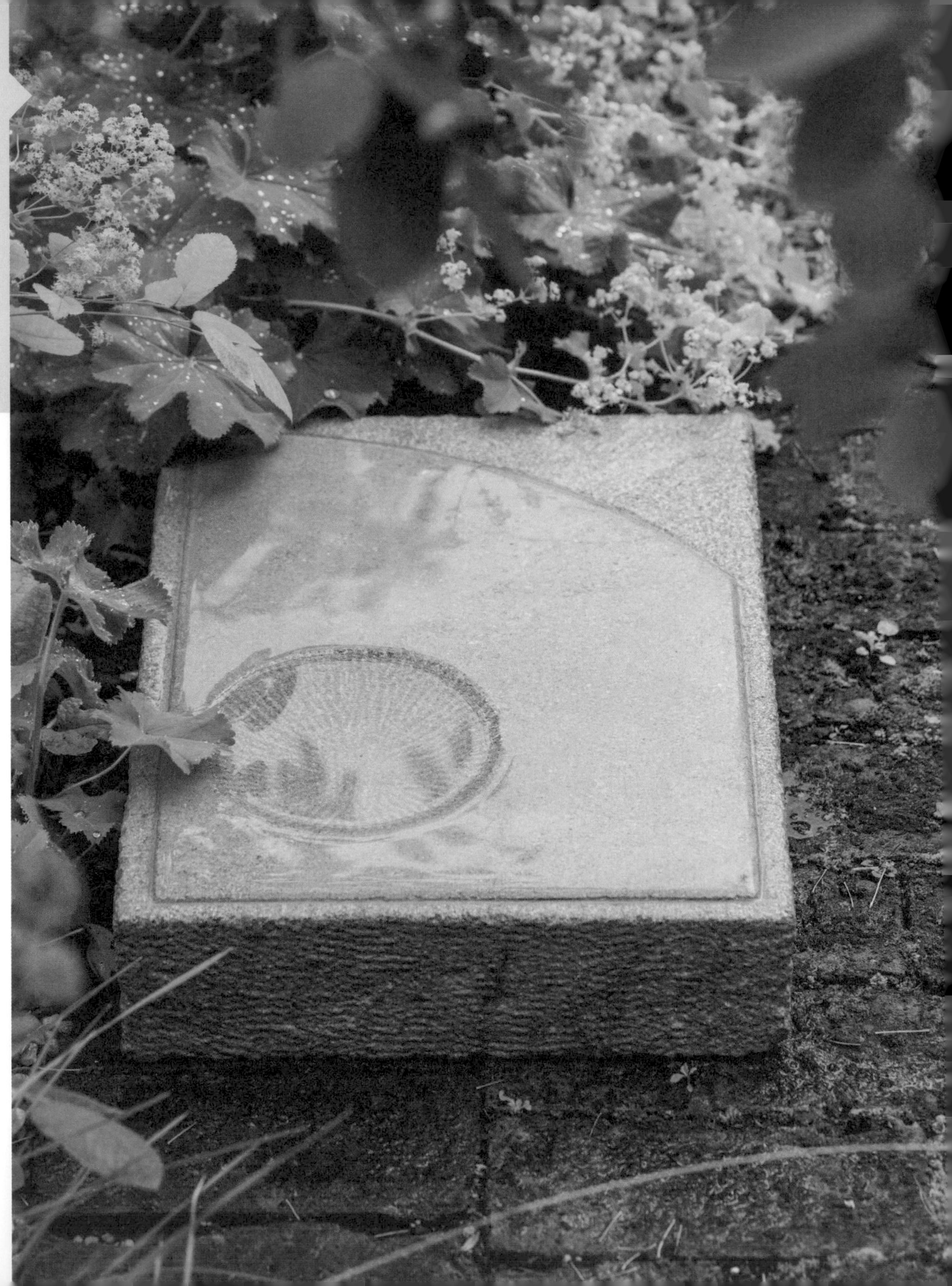

CARVED BIRDBATH

I always garden with wildlife in mind and do whatever I can to encourage it into the garden. Wildlife needs water, and if you can't have any other water feature in your garden, a birdbath is the starting point. Plus, watching birds bathing and other wildlife visiting is wonderful.

YOU WILL NEED

- 17 x 13 x 3½in (430 x 320 x 90mm) slab of limestone or other stone

Tools

- workbench, vice, or clamps
- ruler or straight piece of wood
- pencil
- beam compass
- different types of chisels, such as claw bit holder with 1in (25mm) claw bit, ½in (12mm) gouge chisel
- dummy mallet
- tape measure
- gloves (optional)
- dust mask
- safety goggles
- angle grinder
- diamond cutting blade for natural stone
- diamond polishing hand pad
- drill with polishing disk (optional)

For this birdbath I've used a slab of limestone and chisels to cut out a bathing area in one corner. I've added texture around the slab using a claw chisel, which not only gives more interest but helps the stone look more attractive as it ages. The chiselling is really not that difficult—it just takes a little time and practice. The bathing area is 6in (150mm) across and 1⅜in (35mm) deep.

CONSIDERATIONS

- Be mindful of where you position the birdbath. Ideally, it needs to be in a place where you're going to see it often and where it's easy to keep topped up with water. If you can see it from a window, all the better.

- It's important to work in an open environment with good air flow and wear a dust mask, so that you don't breathe in any dust thrown up by cutting stone.

- If you have a rockery, you may be able to carve a birdbath from the existing stone that you have in your garden.

EXPERT INSIGHTS

- You can experiment with different types of chisels, including curved ones, to give the stone an interesting texture. You could just hollow out the bowl and leave the rest plain, or add patterns around the edge.

- I've used limestone, but there are many other types of stone you could use, such as sandstone, granite, alabaster, and soapstone. Bear in mind that some stones are softer than others and so are easier to carve. Try to choose a stone that's local to your area so it blends in well with your garden.

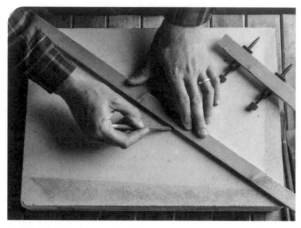

1 Lay your stone on a flat surface at a comfortable working height. You may need to clamp it in place if it moves when you're working on it. Using a ruler or straight piece of timber, mark the stone from corner to corner.

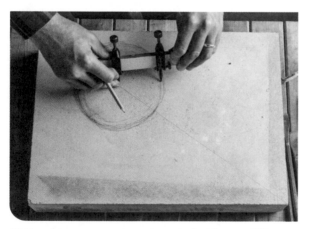

2 Use a beam compass to mark out the bathing area. This can, of course, be placed wherever you want across the stone base, but make sure your circle is at least 1in (25mm) from the edge of the stone. This will allow you to detail your edges without the bath looking too cramped. I have used the diagonal line across the stone as a guide for the center of my design.

3 At the other end of the slab, above the bathing area, use the beam compass to mark a curved line from the top of one side to about a third of the way down the opposite side. This marks the area of the main detailing, which you can leave plain if you prefer.

Small chisels

Although it takes a little time if you use smaller chisels, you're less likely to remove large chunks of stone and so can better control the detail.

> *I like including features with interesting details to catch the eye, or as an element of surprise near a seat or somewhere you're naturally drawn to.*

5 Once the shallow bowl is roughly in place, switch to the claw chisel to texture and fine-tune the design. Try always to work inward, as this will help you avoid taking out large chunks of the stone and losing the straight edge.

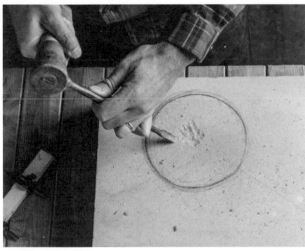

4 Use your gouge chisel to carve out a shallow bowl within the circular shape, making little cuts to get a nice pattern. I tend to keep it quite simple. Here, I'm using a ½in (12mm) gouge chisel and a dummy mallet to take out the center and add texture.

6 Once you're happy with the detailing in the bowl, mark around the perimeter of the stone, ⅝in (15mm) in from the edge. This is where you will add more detail to the stone edges.

7 Carefully work around the edge of the stone with the claw chisel to start the detailing. Take your time, use slow, gentle hammer blows, and keep the chisel very slightly angled into the stone.

9 Use the angle grinder along the marked line at an angle to detail the line. This does take a steady hand. You could clamp a straightedge to the stone to guide you.

8 Next, add some texture around the slab using the same claw chisel and dummy mallet. Work little areas at a time and keep the chisel squarer to the stone than in step 4. This adds interest and helps the stone look better as it ages.

" The chiselling is really not that difficult—it just takes a little time, practice, and patience. "

10 Use a diamond polishing hand pad to smooth off the edging detail.

11 It's not a must, but you could use a drill with a polishing disk to finish off the top surface.

12 When you're putting the bath into position, think about how you can get more from the feature—where you're more likely to see it or walk past it, and how the light will reflect on the water throughout the day.

SPARROW TERRACE

The idea for this bird box was inspired by my love of sparrows and childhood memories of watching them in the garden. You see far fewer of them these days, so I thought it would be good to try to attract more of them into my garden. I reckoned a miniature row of terraced "houses" would suit them well.

YOU WILL NEED

See p104 for timber requirements and measurements

- 1½in (40mm) brass screws
- 1¼in (30mm) galvanized nails
- linseed oil or animal-safe preservative
- 2½in (60mm) screws and anchors for fixing

Tools

- workbench, vice, or clamps
- tape measure
- pencil
- set square
- saw
- cordless drill/screwdriver
- ⅛in (4mm) drill bit and ½in (10mm) countersink drill bit
- 1⅛in (28mm) core drill bit (spade/auger/forstner bit)
- ¼in (6mm) V chisel, flat chisel, and ⅜in (8mm) gouge chisel
- wooden mallet
- small hammer
- brushes or cloth
- level

Rather than build a generic bird box, it's much more interesting to research what birds live in your region and the sort of habitat they are drawn to. It makes the process much more personal, and it's really satisfying seeing the birds you want to attract setting up home in these purpose-built habitats. Like many previously common birds, sparrows have been in decline in highly developed areas of the US for decades but are relatively stable in rural areas. As sparrows generally nest in colonies, I've made this sparrow terrace with four separate entrances as a way of inviting them in. It's made from oak and, depending on what you prefer, you can make the roof out of cedar shingles, copper, or lead.

CONSIDERATIONS

- The entrance hole should be at least 5in (130mm) above the floor of the box to stop young birds accidentally falling out of the hole.

- Leave the inside wall below the entrance hole as rough, unplaned wood, as this helps youngsters clamber up to the hole when it's time for them to leave.

- If you leave the wood untreated, it will gradually turn silver, which helps it to blend into the setting. However, softwoods can be treated with an animal-safe, water-based preservative. Apply it only to the outside of the box and avoid using it around the entrance hole.

EXPERT *INSIGHT*

- The roof overhangs the front of the box—this will help to keep rainwater off and make the box watertight.

- Position the box between north and east to avoid it being in strong sun. It should be placed 6½–13ft (2–4m) high on a tree or wall.

- The best time to put up bird boxes is in early fall as many birds are looking for places to roost at that time. They may use the same box for nesting in spring.

SPARROW TERRACE PIECES

Use the pattern below as a guide to measure, mark, and cut your timber to size. You will need 1in (25mm) thick timber, such as air-dried oak.

FRONT VIEW

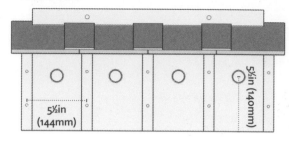

5½in (144mm)

5½in (140mm)

SIDE PANELS x 2

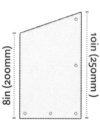

8in (200mm)

10in (250mm)

5¾in (145mm)

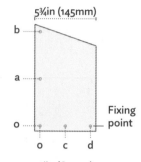

b

a

o

o c d

Fixing point

o–a = 4⅓in (110mm)
o–b = 8¾in (220mm)

o–c = 2½in (60mm)
o–d = 4¾in (120mm)

FRONT PANEL

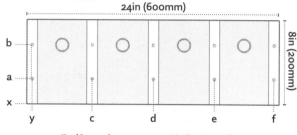

24in (600mm)

8in (200mm)

b

a

x

y c d e f

y–a = 2⅓in (60mm)
y–b = 5½in (140mm)
x–y = ½in (12.5mm)

x–c = 6in (156.5mm)
x–d = 11¾in (300.5mm)
x–e = 17½in (444.5mm)
x–f = 23in (588.5mm)

CEDAR SHINGLES FOR ROOF x 7
⅝in (15mm) thick

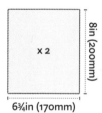

8in (200mm)

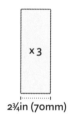

x 2

x 2

x 3

6¾in (170mm)

6in (150mm)

2¾in (70mm)

BACK PANELS x 2

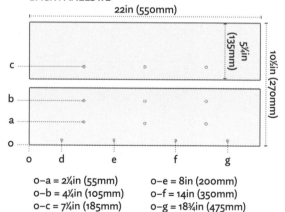

22in (550mm)

5¼in (135mm)

10½in (270mm)

c

b

a

o

o d e f g

o–a = 2¼in (55mm)
o–b = 4¼in (105mm)
o–c = 7¼in (185mm)
o–d = 3in (75mm)

o–e = 8in (200mm)
o–f = 14in (350mm)
o–g = 18¾in (475mm)

INSERT PANELS x 3

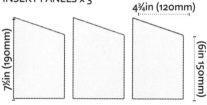

7½in (190mm)

4¾in (120mm)

(6in (150mm))

ROOF PANEL

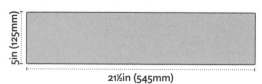

5in (125mm)

21½in (545mm)

BASE

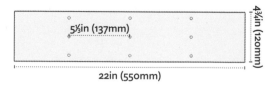

5½in (137mm)

4¾in (120mm)

22in (550mm)

1 After cutting your timber, offer up all your pieces to understand how it sits together. Double-check all the measurements and add pencil marks as needed as guides.

3 As you're using hardwood, it's a good idea to pilot drill all your holes after you have marked up all your timber. Alternatively, you can pilot drill as you work, whatever you find easier. If doing so now, pilot drill your fixing holes using a ⅛in (4mm) drill bit.

4 Then use a ½in (10mm) countersink to drill all of the exterior-facing holes so that all the screws will sit flush.

2 Once you have put everything in place, measure and mark all your fixing points and the position of the compartments with a pencil and set square, using the plans opposite as a guide.

> " *You can discover lots of information about birds on the National Audubon Society website, or do some research to find out about local birdwatching groups.* "

5 To mark the entrance holes, lay your front panel on a flat surface. At the center of each compartment, draw a vertical line. Measuring from the bottom of the front panel, make a mark 5½in (140mm) up on the vertical line. This will be the center of your entrance holes.

What size?

Adapt the size of the entrance hole according to the bird species you'd like to attract:

○ 1–1⅛in (25–29mm) for Carolina wren, house wren

○ 1¼in (32mm) for chickadee, tufted titmouse, downy woodpecker

○ 1½in (38mm) for Eastern bluebird, tree swallow, hairy woodpecker

○ 1¾–2in (45–51mm) for red-bellied woodpecker, flycatcher

6 Clamp your boards to a flat surface. Then, using a 1¼in (28mm) core drill bit, drill out the four entrance holes. These holes are the right size for tree sparrows and some other birds, but you'll need to make the holes smaller or larger for certain other species (see box above).

7 If you haven't already done so, mark and pilot drill your fixing holes across the front panel at 2⅜in (60mm) and 5½in (140mm) from the bottom center of the area that will have the detailing.

This project will be easier to manage if you have someone helping you. Fixing the components together can be a bit fiddly.

8 First, use the V chisel to work along your two pencil lines. Once you have marked both sides, use the flat chisel to deepen the cut (see p33 for more detail). This will give you a good edge to work your gouge chisel into.

9 On the front panel, use a ⅜in (8mm) gouge chisel to create your detailing. Once you have an angled cut, work the chisel from the middle out (see p33 for more detail). If you find it difficult to use the chisel, you could use a wooden mallet to help.

10 Using the ⅜in (8mm) gouge chisel, work slowly around the outside edge of the entrance holes, removing small slivers of wood. I work the chisel at an upright angle and use the mallet rather than pushing the chisels. Just work slowly around the edges.

12 Before you start fixing the front and roof, fix the insert panels into place with 1½in (40mm) brass screws. Fix the screws partly through the predrilled back panels so you can just see the screw. This will help you align your dividers.

11 You can now start fixing your box together. Lay the base on your work surface, and then put the two sides into place, fixing with 1½in (40mm) brass screws as you go. Then fix the two back panels to the sides and base.

> " *If you've never used a chisel before, it's a good idea to practice on a few pieces of old wood until you feel confident. It's not difficult, but it takes time.* "

13 Once the front panel is fully detailed, fix it in place with 1½in (40mm) brass screws. Then put the roof panel in place, making sure the angle of the roof lines up with the side panels.

14 Measure and mark up the cedar shingles for the roof.

15 Cut the cedar shingles to size, using a workbench or clamps to hold the wood firmly in place.

16 Put your roof shingles in place. The 6¾in (170mm) shingles go in the center, with the 6in (150mm) shingles on the two outside edges to create an overhang. The 2¾in (70mm) shingles will cover the joints between the bigger ones. Before fixing the shingles, draw a pencil line along the top, 2in (50mm) down from the back board. This is to guide the position of the nails.

18 Put the small shingles in place so that they cover the joints in between the bigger shingles and will keep rain out. Nail into place with two fixings at the top and at the bottom of each shingle.

17 Fix the roof tiles into place with 1¼in (30mm) galvanized nails, using your pencil line as a guide.

19 Use a cloth to rub the outside surface with linseed oil. Make sure you dispose of the cloth carefully afterward as rags soaked in linseed can ignite. Alternatively, you could use an animal-safe preservative on the surface.

20 Work out where best to position your nest box. Ideally, it will be near a supply of food and materials that birds might use for nesting. Measure and mark the fixing points, using a level and a tape measure to ensure the box will sit level.

22 The chiselled detailing not only adds visual interest but also helps camouflage the screws. The detail really does improve with time.

21 Fix the nest box in place using 2⅜in (60mm) screws and anchors. I've put mine on a wall in a semi-shady position.

Fixing

Attaching a bird box to a tree with metal nails or screws can cause damage to the tree, so it's better to attach it with wire covered in a length of hose or bicycle tire around the trunk or branch. Alternatively, use a nylon bolt.

TOAD HOUSE

When I'm designing a garden, I like to create habitats to encourage wildlife into it. I find out what sort of conditions certain creatures look for and then try to recreate them. Toads like dark, damp, tucked-away places such as under stones or wood, so I created this little house to look a bit like a textured boulder.

YOU WILL NEED

- block of green oak 8 x 6¼in (200 x 160mm)

Tools

- pencil
- saw
- workbench, vice, or clamps
- drill and 1in (25mm) drill bit
- ¼in (6mm), ½in (10mm), and ¾in (20mm) gouge chisels
- wooden mallet
- sharpening stone (for chisels)
- sandpaper

I came up with the idea of this little carved house for one of my Chelsea Flower Show gardens in 2013. It's made from an offcut of an oak sleeper and measures around 7 by 5½in (180 by 140mm) across, with a little notch for an entrance. It's quite flat so it sits firmly on the soil and its subtle sculptural quality helps add a touch of interest to a quiet corner. You could use most other sustainable hardwoods, such as maple or cherry. Toads and frogs are great amphibians to have in the garden as they eat lots of different pests such as caterpillars, slugs, bugs, and cutworms.

CONSIDERATIONS

- For this project, it's a good idea to kit yourself out with a mix of different chisels to work with, and you can use them for a number of other projects in this book too (see pp32–33).

- This is a great little project to practice your carving technique as it's really forgiving.

- I tend to leave this looking natural and don't treat it with oil, and I use a hardwood so it will last longer.

EXPERT INSIGHT

- Toads have dry, bumpy skin and live most of the time on land. They vary in color from green to olive or dark brown, sometimes gray. They can grow up to 5in (13cm) long and the females are larger than the males. Unlike frogs, they walk or crawl rather than hop.

- Toads particularly love damp places like piles of leaf litter, logs, or stones. Providing them with some shelter means they can keep out of harm's way and be protected from frost.

3 Hold your block of wood firmly in place with a vice, clamps, or a workbench. To make chiselling the inside easier (see step 6), drill several large holes to remove a decent amount of wood. You can mark the depth on the drill bit so you don't go too deep. Be careful when you start drilling as the larger drill bit can bite a little.

1 Use a pencil to mark out the shape of your toad house on the wooden block.

2 Saw off the corners to make the block as rounded as possible.

4 Reposition the block so you can work on the outside. Using the chisels, work your way around the outside surface to get the rough shape you want to achieve. It's important to carve along the grain line. You can carve diagonally along the line, but don't carve up against the grain as you might find that the wood starts to tear.

5 Use the various different-sized chisels to get the texture and pattern you're aiming for. Make sure you keep your chisels sharp with a sharpening stone. Keep moving the block to aid your carving and keep working down along the grain.

6 Flip the block over and secure it again. Use the chisels to carve out the inside and create a hollow.

7 Finally, make an entrance for the toad by cutting a curved notch into the rim about 1½–2in (40–50mm) wide and 1¼–1½in (30–40mm) at the highest point. When you're happy with the results, use some sandpaper to smooth off any sharp edges. Then find a damp spot in your garden to put your new toad abode.

" *The key to this project is getting a set of different-sized chisels, and then it's just a matter of playing and having fun creating interesting textures on the outside surface of the wood.* "

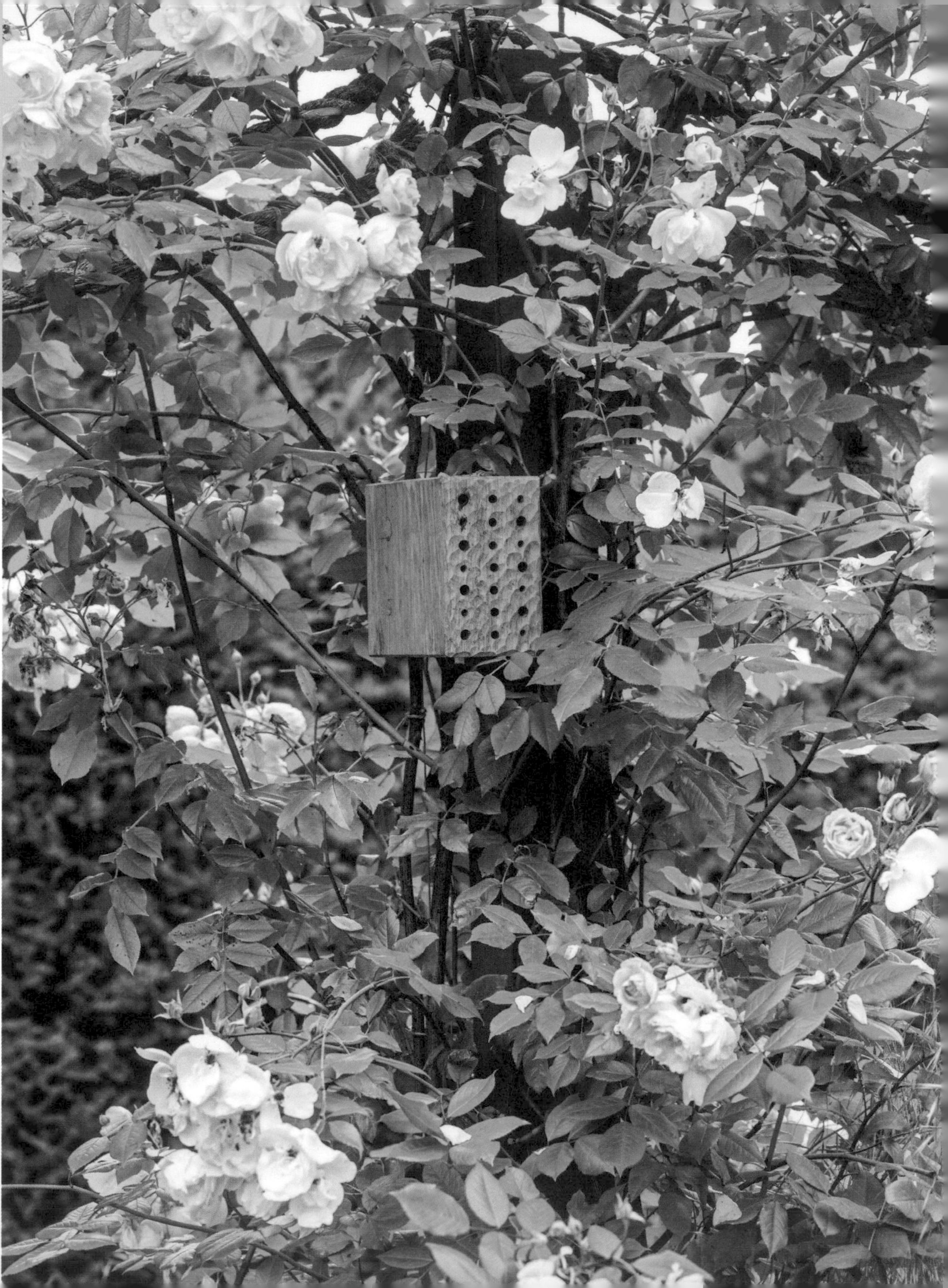

BEE HABITAT

Is there anything more wonderful in a garden than watching a busy bee at work? Bee populations have been declining for many years, so I wanted to create a little habitat to encourage them into my garden. The block is an offcut from an untreated oak sleeper and I've drilled lots of holes in it to look like tiny windows.

YOU WILL NEED

- block of timber, such as oak or any hardwood, cut to 4in (100mm) wide x 6¼in (160mm) high x 8in (200mm) deep
- 2¾–4in (70–100mm) screws for fixing
- **Tools**
- wood planer
- workbench, vice, or clamps
- hand saw, chop (abrasive) saw, or electric circular saw
- set square
- pencil
- tape measure
- safety goggles
- cordless drill/screwdriver
- drill and mix of drill bits, such as ¼in (6mm), ⅜in (8mm), and ½in (10mm)
- wooden mallet
- ¼in (6mm) gouge chisel
- sandpaper
- ½in (10mm) drill bits for countersinking timber locks and ¼in (6mm) for pilot holes

Bees have been around for about 130 million years, but they are now in serious decline, mainly because of climate change, loss of habitat, parasites, and the use of pesticides. We can help them out by providing pollen-rich plants for food and somewhere safe for them to live. With this bee habitat, I did lots of research and was aiming to attract mason bees, but it was leafcutter bees that moved in. I reckon when they saw the roses on their house search, they thought "job done!"

CONSIDERATIONS

- It's a good idea to check it and clean it out ready for the next year. I tend to clean these habitats in situ once the bees have left their home— holes that are covered with leaves or mud may still be in use. Use warm water or a pipe cleaner to brush out any debris.

- Rather like people, bees all want different things. Some live in shells, others live in the ground, and some in trees, but most of our common species are just looking for a dry, dark cavity in which to live.

- If we're gardening with wildlife in mind, it's good to create habitats, but they can be beautiful as well as useful.

- I've used a core drill bit so that I can countersink the screws for fixing.

EXPERT INSIGHT

- You could use various different woods including softwoods. It would take a long time for a block like this to rot away. The key is that it should be untreated and from a sustainable source.

- Make sure the habitat is in full sun with a clear access in and out. Once in place, you can surround it with plants for pollinators. Don't just think about summer flowers as most bees will be looking to set up home in spring. Plants that flower early in the year, such as crocus, cyclamen, snowdrops, and hellebores, are important too.

1 If the timber you're working with is a bit rough, it might be worth running a planer over it so that you have a clean, smooth block to work with.

3 Using a set square, mark three straight lines lengthwise on the front of your block. Start with the center line, then mark a line either side of that centered between it and the edge of the block. Starting ¾in (20mm) from the top, mark across the lines every 1⅝in (40mm). These are the drill marks for your holes.

2 Cut your timber to size, ideally using an electric circular saw or chop saw if you have one. If not, a hand saw is fine.

4 Secure your piece of wood firmly and put on safety goggles before drilling. Use long drill bits to drill into the wood as far as you can so that the bee can use the whole cavity. I used ¼in (6mm), ⅜in (8mm), and ½in (10mm) drill bits to create different-sized holes.

7 On the side where you will attach your bee house, use a set square and pencil to mark where you want to put your fixings. Drill two holes to create a cavity to countersink your fixings. Next, pilot drill the rest of the fixing holes with a ¼in (6mm) drill bit—this will make fixing easier.

5 Use a wooden mallet and a chisel to create some interesting textures on the surface of the wood around the holes.

6 Tidy up the entrance holes with a piece of sandpaper wrapped around a pencil or a large nail to remove any splinters.

8 Attach the bee habitat with a screw through the fixing holes to an arch post, fence post, gate post, or a wall. I've fixed my habitat onto a vertical post near some planting. The main thing is to make sure it's in a sunny position near lots of pollen-rich plants.

INSECT SEAT

This design takes a simple bench to another level of usefulness. Not only is it a place to sit and use as a focal point in the garden, but it's also a great habitat for beneficial insects. I made this one from oak and infilled the void beneath the seat with different-sized logs I'd collected from around the garden.

YOU WILL NEED

See p122 for timber requirements and measurements

- ○ 2¾in and 3in (70mm and 80mm) screws
- ○ ¾–7in (20–180mm) diameter logs in a mixture of lengths

Tools

- ○ workbench, vice, or clamps
- ○ pencil
- ○ tape measure
- ○ set square
- ○ handsaw
- ○ hand planer
- ○ 2–3in (5–8mm) wooden spacers
- ○ set square
- ○ power drill
- ○ ⅛in (4mm) drill bit and ⅖in (10mm) countersink bit
- ○ cordless drill/screwdriver
- ○ lawn edger
- ○ spade
- ○ level
- ○ trowel
- ○ hand sledge
- ○ bow saw or chainsaw

This bench is a pretty straightforward structure. Essentially three long lengths of oak form the seat with three shorter lengths for the uprights at each end. A crossbar and fixing bars hold everything in place. I set it into the ground so it's nice and stable, then I infilled the space beneath the seat with logs of various diameters cut to length. These form lots of lovely little niches where insects and other critters can tuck themselves away. In time, the logs will rot down, but they can easily be pulled out and replaced. I love the idea of insects and creepy-crawlies going about their business inside it.

CONSIDERATIONS

- It's best to place this in a relatively shady spot that's not exposed and with plenty of shelter. You want to encourage earwigs and other creepy-crawlies to live there, so you don't want it overheating. And if it's close to water, so much the better. The added benefit of being near water is that as well as insects, you may get a toad moving in.

- Think about its position from a design point of view too. Do you want to add something to a scene, such as tucking it back in a border, putting it on the edge of woodland, or by a pond?

- Use logs that look attractive both at the front and the back of the seat, so you can use it on both sides.

EXPERT INSIGHT

- ○ Oak, being a hardwood, will last a long time without being treated. It silvers as it ages and will possibly twist a bit, but it's a good long-lasting material. You could also use treated softwood.

- ○ It's best to use an array of diameters for the logs, not only because they give more visual interest but also because the varying sizes of nooks and crannies suit different creatures.

- ○ You can vary the height of the bench—you could make a smaller one for kids to sit on, for instance.

TIMBER FOR THE INSECT SEAT

Use the measurements below to measure, mark, and cut your timber to size. You will need 2¾in (70mm) thick oak planks and 2 x 2in (50 x 50mm) sawn treated timber for the framework.

UNDERSIDE OF SEAT

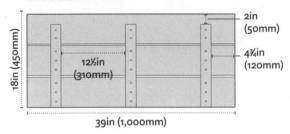

18in (450mm)
2in (50mm)
12⅕in (310mm)
4¾in (120mm)
39in (1,000mm)

SEAT PANELS x 3
Oak planks

5½in (140mm)
39in (1,000mm)

SIDE PANELS x 6
Oak planks

5½in (140mm)
24in (600mm)

FIXING BARS x 5
Sawn treated timber

2in (50mm)
16in (400mm)

CROSSBAR
Sawn treated timber

2in (50mm)
31in (770mm)

FRONT/BACK OF SEAT

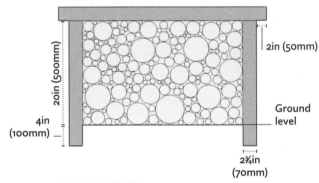

20in (500mm)
2in (50mm)
4in (100mm)
Ground level
2¾in (70mm)

INTERNAL FRAME

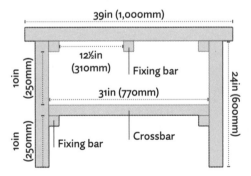

39in (1,000mm)
10in (250mm)
12⅕in (310mm)
Fixing bar
24in (600mm)
31in (770mm)
10in (250mm)
Fixing bar
Crossbar

⚠ CAUTION

If you decide to cut your logs with a chainsaw, be aware that it is one of the most dangerous power tools around. Make sure you know how to use it safely before cutting anything and always wear protective boots, gloves, and safety goggles. Also make sure there is another person with you, you have a first aid kit close to hand, and you can get to the nearest hospital quickly if need be.

SIDE PANELS

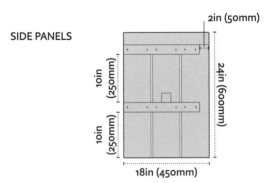

2in (50mm)
10in (250mm)
24in (600mm)
10in (250mm)
18in (450mm)

1 Position your lengths of timber on a workbench, and then use a pencil, tape measure, and set square to mark all your timber to size according to the measurements on the opposite page. It's a good idea to mark your timber on three sides.

3 Using a hand planer, take off the corners and edges of all pieces of timber, aiming for a 45-degree angle. This gives a neater, softer look and helps to create lovely shadow lines down the sides of the uprights. Plane all outward-facing wood until it is smooth to the touch.

2 Carefully cut the timber to size using a handsaw, making sure you follow the pencil lines you have drawn. This essentially gives you a "kit" to put together.

4 Lay your seat timbers on a flat surface. Put wooden spacers in between the lengths at either end, making sure everything is square and aligned. Mark a line where the inside of the side panels will be fixed by measuring 4¾in (120mm) in from the edge of the wood on both sides. I've designed this so that there is a 2in (50mm) overhang on either side.

6 Pre-drill all the marked holes with a ⅛in (4mm) drill bit, ready for the fixings.

7 Countersink the holes you made in step 6 with a ½in (10mm) countersink bit.

5 Lay out three fixing bars, making sure the ends are flush. Position the first one so that it lines up with the pencil mark made in step 4 and sits equidistant from the sides of the seat timbers. Mark a line across them to indicate where to pilot drill the fixings. You need six evenly spaced holes and the screws should go neatly into the seat timbers.

9 Repeat the process with the side panels. Set the fixing bars 2in (50mm) back from the front with the top edge 12in (300mm) below the top of the side panels.

10 Put the seat timber on the workbench with the fixing bars uppermost. Put one of the side panels in position, making sure it's upright and everything is square. Use 3in (80mm) screws to fix the side panel in place.

8 Align one of the fixing bars to the pencil mark on the seat panels you made in step 4. Set it 2in (50mm) back from the front and flush with the back. Use 2¾in (70mm) screws to fix it to the seat. Repeat the process with the other two bars, fixing the second one at the other end of the seat and the third one in the middle.

11 Repeat the process with the second side panel, making sure everything is butted up tightly. Once the two sides are fixed, turn the seat upright. Put the crossbar in place and screw it into the center of the fixing bar on the side with 3in (80mm) screws. The crossbar acts as a control to stop the sides wobbling and kicking out.

12 Decide where you want to position your bench. If you want it in a grassy area, you'll need to remove the sod. Use a lawn edger to mark out the area you need to dig.

13 Put the seat aside to give yourself room to move, and dig out enough soil so that you can sink the sides 4–6in (100–150mm) into the ground. This will help to stop the bench from moving around. Once you've dug out the site, roughly level it off.

> ❝ I really like the idea of creating simple, easy-to-make features for the garden that are both beautiful and useful. ❞

15 Start putting your selection of logs in place. The first layer sits below ground level. This helps hold things together and creates more habitat for creatures that live in the soil.

14 Drop the bench into place and check everything is level and square with a level. Back-fill soil around the legs and ram it firmly in place with a hand sledge. If you haven't already cut them, cut your logs to size with a bow saw or a chainsaw.

16 Continue adding layers of different-sized logs beneath the seat, slotting smaller logs into the gaps so the whole space is filled.

PROJECTS

FURNITURE

> *Having things around the garden that mean something to you will always bring a smile.*

INTRODUCTION

Out of all the projects we've included in the book, I reckon those included in the furniture section will give you the best opportunity to personalize your space. And although some of the projects here take a certain amount of skill, I think we've shown that they are all really doable.

With furniture, it seems that most people generally choose to either go cheap and cheerful or opt for something more classic and expensive. Either way, it's all too easy to have furniture that looks like everyone else's. These projects, on the other hand, give you the scope to have a bigger impact on the spaces you're creating. The ideas behind them all came out of a memory, a particular time or place, which means I feel a very personal connection to them. As such, they have my style and personality stamped all over them, but you can easily give them your own style— whether it's choosing different colors or materials, or adding other elements of interest such as textural detail—and all of this will help to make your garden feel uniquely yours.

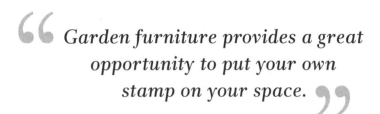

Garden furniture provides a great opportunity to put your own stamp on your space.

COFFEE TABLE PLANTER

The idea with this project is to make a living coffee table so that you can enjoy the plants close up while you're having a drink. It's made out of an old oval galvanized bath that I've filled with succulents to create a miniature undulating landscape that will really draw you in.

YOU WILL NEED

- 2 pieces of stone about 4 x 4¾in (100 x 120mm) and 1–2in (25–50mm) thick
- galvanized bath 25in (630mm) wide x 32in (800mm) long x 14in (350mm) deep, or other galvanized container
- crocks
- gravel
- cactus potting mix
- low-growing plants, such as succulents

Tools

- pencil
- metal ruler
- vice, clamps, or workbench
- angle grinder, safety goggles, and dust mask
- claw bit chisel with 1in (25mm) claw bit
- small hammer or dummy mallet
- cordless drill and ¼in (6mm) drill bits for metal
- trug
- trowel

Succulents I used in this project

- *Sempervivum allionii*; *S. arachnoideum*; *S. arachnoideum* 'Rubin'; *S. calcareum*; *S. calcareum* 'Guillaumes'; *S.* 'Green Ice'; *S.* 'Irazu'; *S.* 'Mulberry Wine'; *S.* 'Pacific Purple Shadows'; *S.* 'Red Delta'; *S.* 'Reinhard'

I've used various tones of succulents to offset the gray of the container. Among the planting, I've set a couple of flat pieces of stone, like place mats, so you can perch your drink on them. The container has handles so you can pick the whole thing up and move it to wherever you fancy sitting, whether it's for your morning coffee or drink in the evening, but you could make a smaller one using a bucket.

CONSIDERATIONS

- The idea of getting close to living things, so you can enjoy and appreciate them, was the inspiration for this project. .

- What I love about succulents is their sculptural forms, which can make beautiful patterns when planted together. Also, once you've planted them, they don't need much looking after. Just make sure you add a layer of grit to the surface of the compost to stop the leaves from becoming too wet and rotting.

- You can use herbs or other low-growing plants instead of succulents if you'd prefer. Whatever you choose, use colors that will work well with the color of your container. You can make the planter as big or as small as you'd like.

EXPERT *INSIGHT*

- I recommend using stone for the place mats that's around 1–2in (25–50mm) thick so you can chisel a little texture into it.

- I've used a couple of pieces of limestone because it's local to my area, but you can use whatever you have lying around in your garden. You could use sandstone, brick, chunks of paving slab, or a tile, for instance.

1 Use a pencil to mark out the shape you want your place mats to be. I have marked out a trapezium shape measuring about 4 x 4¾in (100 x 120mm), but you can do any shape you like as long as you can fit a mug or glass on it.

3 Once you've cut your shapes, work the surface of the stone with a claw bit chisel and dummy mallet to get the effect you want. I used the claw bit in different directions to create an interesting texture. The surface of the stone will weather beautifully over time.

2 Secure the stone firmly on the ground so it won't move. I sometimes wedge it between larger pieces of stone. Then use a small angle grinder to cut out your shape. Or use a masonry chisel and hand sledge. Remember to always wear safety goggles and a dust mask when you're cutting stone.

" *I love using succulents because they are so sculptural, and they create such dynamic patterns when you plant them closely together.* "

4 Work the edges carefully to give them a softer look.

Containers
I love using these old bathtubs, but anything of a similar size could work. The handles are useful for picking it up and moving it around, but bear in mind it could be heavy.

6 Turn the bath over again and put crocks (broken pieces of terracotta) across the bottom of the container to aid drainage.

5 Turn the bath upside down and drill a series of ¼in (6mm) holes across the base. This is really important for drainage. This bath has a little rim on the bottom so the base sits just off the ground. You could put some bricks, stone, or timber underneath to lift the container slightly if you're concerned about drainage.

7 Then add a layer of gravel on top of the crocks, which will also aid drainage.

9 Position your stone placemats in the soil on either side of the container. Make sure they are level and sitting firmly. Don't put them too close to the edge as you want the plants to wrap around them.

8 Next, using your trug, make up a planting mix of roughly 60 percent potting mix and 40 percent fine gravel. Add the mix to the container, allowing enough space at the top to add your plants. Make sure the plants will sit just below the rim—this will help to contain your gravel dressing.

Plant choice

Whatever plants you choose, it's a good idea to pick a range of colors that will work well with the color of your container.

> *For me, it's all about creating those quiet little moments when suddenly you find yourself completely transfixed because you're reminded just how fascinating nature is.*

10 Next, arrange your plants around the place mats. I start by putting in a few key plants, and then pick out one color and use other plants to create a rhythm. It's a bit like creating a tapestry. Make sure that there are plenty of undulations in it.

11 Then all you need to do is top-dress with gravel to make the arrangement look smart. Gravel also helps to keep moisture away from the base of the plants.

12 Finally, position your finished coffee table planter in a suitable spot in the garden. Ideally, it will go in a place that is light and spacious enough to be able to pull up a couple of comfortable chairs around it.

SIMPLE LOG SEAT

I love the idea of perching on a log in a woodland setting, just enjoying the surroundings. These seats are easy to make as they are simply tree trunks cut to a suitable height. The key is to make them look natural and put them in the perfect spot to enjoy the view or the beauty of the light at a certain time of day.

YOU WILL NEED

- logs about 20–24in (500–600mm) high x 12–16in (300–400mm) wide
- linseed oil or marine oil

Tools

- tape measure
- chainsaw (if you cut your own logs)
- ⅝in (14mm) and ¾in (20mm) gouge chisels (optional)
- hammer or mallet
- drill and drill bits (optional)
- sandpaper (optional)
- cloth (to apply oil)
- spade or shovel
- level
- hand sledge

CAUTION

If you cut your own logs, be aware that a chainsaw is one of the most dangerous power tools around. Make sure you know how to use it safely before cutting anything and always wear protective boots, gloves, and safety goggles. Ensure there is another person with you, you have a first aid kit close to hand, and you can get to the nearest hospital quickly if need be.

Think of these seats as being a destination, somewhere you'd like to go to and sit with a cup of tea. Find a place where you can appreciate the simple beauty of nature—somewhere tucked away in dappled shade with a view would be perfect. They look particularly good in a woodland setting. What I really love about this log seating is that, over time, the logs will ever so gradually rot down, then disappear back into the earth, leaving nothing behind.

CONSIDERATIONS

- The look you create depends on the dimensions and type of logs you use. Make sure you choose timber that does not seep. I tend to opt for a hardwood, such as oak, walnut, or cherry.

- Unless you're proficient at using a chainsaw, it's best to ask a timber merchant to cut the logs to your preferred height. A local sawmill or tree surgeon might also be able to supply the cut logs.

- I've put my log seats in a woodland setting and interplanted ferns around them. I love looking down on the small detail of plants while I'm perching on one of these seats.

EXPERT *INSIGHT*

- These are really simple seats, and I've made them a bit more interesting by chiselling detail into the top, but you can leave them plain if you prefer.

- Try and personalize everything you put in the garden to make it yours. If you live with children, you might want to do smaller log seats especially for them.

3 Then I used a ¾in (20mm) gouge chisel to detail the outside edge of the seat, working from the bark inward.

1 Work out where you want to put the seats, making sure that it's a comfortable place to sit with something attractive to look at. Measure the height you want the seats to be. The best way is to take the measurement from the ground behind your foot to the back of your knee. The average seat is about 18 to 20in (450 to 500mm) high. Add on another 4in (100mm), which will be dropped into the ground to keep the seat stable.

2 If you'd like to add detailing to the top, it's best to do it at this point. Here, I used a ⅝in (14mm) gouge chisel. Starting at the center, I worked my way out using the growth rings as a guide.

4 Use a cloth to apply two coats of marine oil or linseed oil to the textured surface of the seat. Apply liberally and let the oil soak right into the wood. It seals it against water and really brings the timber alive. Allow to dry completely between coats.

5 Dig a hole 4in (100mm) deep and slightly wider than the log you intend to use. Use a level to check that the base of the hole is as level as possible. Drop your log into position and use the level across the top in both directions to make sure it's level. If necessary, use pieces of stone and soil to make sure the seat is level and upright.

6 Use a hand sledge to ram the soil back in around the log so everything is held firmly in place. You don't need to use concrete because the logs are quite heavy anyway.

Detailing

There are lots of different ways to add in a touch of detail and personalize this seating, such as drilling holes for insect habitats, chiselling detail into the top, or sanding the tops smooth so you can see the growth rings.

7 Surrounded by ferns, a log seat is the perfect place to observe the small plant details in my garden.

RESTORING FURNITURE

When it comes to garden furniture, you have a few options. You can either spend a small fortune, buy cheap and cheerful stuff, or restore or repurpose older pieces, which is better for the environment. I particularly love these chairs as they remind me of long family lunches on holiday in France.

YOU WILL NEED

- liquid paint stripper
- iron oxide primer for metalwork
- paint thinner
- replacement wood (if needed)
- linseed paint
- raw linseed oil
- balsam turpentine
- wood filler
- replacement screws, either zinc or brass (use the existing ones as templates)

Tools

- workbench (optional)
- hammer (useful for dismantling)
- screwdriver, hand and cordless
- adjustable wrench and pliers
- electric heated paint stripper
- metal scraper
- steel wool
- sandpaper
- paintbrushes
- tape measure
- pencil
- set square
- saw
- hand planer
- cordless drill
- cloths

These days, I'm much more interested in this last approach. I love the idea of things living on and being part of that journey. Vintage pieces invariably have more character too. We found this table and chairs online, but you can find great stuff in antique shops, at garage sales, on Freecycle, and even in dumpsters. You just need to work out how you can adapt, paint, and personalize it to suit the look and atmosphere of your garden.

CONSIDERATIONS

- I've used Brouns & Co.'s linseed paint in Highfield Green, Olive, Maastricht Blue, and Fairburn for this project. It's made from natural ingredients, and although it's a little more work than regular paint because you need to build it up in layers, it's extremely long lasting, lovely to work with, and has really good coverage.

- For wood, you'll need a primer coat of 50 percent paint, 35 percent raw linseed oil, and 15 percent balsam turpentine. Measure out the ingredients and give the paint a good stir, then mix everything together well. After you've applied the primer and allowed it to dry for a day or so, you can then use paint straight from the tin for the subsequent coats. It's normal for the primer coat to look a bit patchy after it's dried. The next two coats of paint will sort that out.

EXPERT INSIGHT

- You don't have to stick to one color—you can mix and match. It's interesting to look at color charts to see how things might go together.

- In between coats, put the brushes in a pot of raw linseed oil to prevent them drying out. To clean brushes, use paint thinner or a mild detergent and warm water, rinsing until all paint has been removed.

- PLEASE NOTE that balsam turpentine is flammable, and rags soaked in linseed paint can spontaneously combust, so make sure you soak any cloths in water before disposing of them.

3 Use steel wool on any metal surfaces to smooth everything down and remove any flaking metal. Use sandpaper for wooden slats to get everything smooth.

1 First, dismantle the furniture, undoing bolts and screws and separating the wooden and metal parts. A hammer, screwdriver, adjustable wrench, and pliers can be useful. Lay everything out on the bench and don't throw anything away in case you need it later. Check if any wooden slats need replacing.

2 Next, you need to strip off any paint. Use the heated paint stripper and a scraper for as much of the work as possible. If there are any awkward bits, such as where parts have been fixed together, then use the liquid paint stripper (following the instructions on the tin).

4 If you're restoring metal furniture, protect the metal from rusting by applying a couple of layers of iron oxide primer, allowing it to dry well between each coat. Keep the brushes in paint thinner between coats to stop them drying out.

5 Give the metal parts two coats of linseed oil paint, making sure that the first coat is thoroughly dry before applying the second.

6 If any of the wood needs replacing, find a wood that's close to it in thickness and quality. Measure the original or use it as a template to draw around. Check the edges are straight with a set square, then cut the wood using a saw. Fill any holes with wood filler, leave to dry, and plane and sand the wood smooth.

7 Mark then drill out holes for screws. Apply a coat of linseed primer to the wooden slats (see p143). Allow it to dry thoroughly, which usually takes about a day. Using the linseed oil paint straight from the tin, or loosened with a little oil, apply a thin layer to coat the wood evenly. Allow to dry thoroughly, then apply a second top coat.

8 Once everything is dry, put it all back together again. You might need to replace some screws, so try to get like for like. To make sure the screws fit tightly into the existing wood, it's often better to use slightly thicker screws.

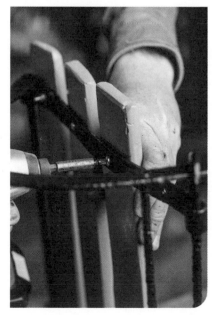

" I love finding stuff that already has a history to it, then restoring it and personalizing it with color. I find the whole process a joy and really satisfying. "

LOW BENCH

This bench was inspired by my visits to Japan.
I love the way lots of Japanese seating is very low
down and how this changes your sightline. When
I started making it, pretty much everyone said
"It's a bit low isn't it?" Then they sat on it and all
agreed that the height was perfect.

YOU WILL NEED

- 2 x 4 x 8in (100 x 200mm) crossties cut to 6ft (1,800mm) lengths
- 2 x 4 x 8 x 14½in (100 x 200 x 360mm) crossties
- 8 x 4in (100mm) or larger exterior wood screws
- wood glue
- 8 x ½in (12mm) oak plugs

Tools

- set square or carpenter's square
- tape measure
- pencil
- cordless circular saw or handsaw
- hand planer
- belt sander or sandpaper
- selection of gouge chisels, including ½, ⅝, ¾, 1½in (10, 14, 20, and 40mm), and curved gouge chisel
- carver's mallet
- spokeshave or spoke plane
- piece of timber 4 x ¾in (100 x 20mm)
- ¼in (6mm) spacers
- ½in (12mm) core drill bit for countersinking screws
- cordless drill/screwdriver
- ¼in (6mm) drill bits
- wooden mallet
- cardboard for protecting timber

The seat is really simple to put together. Essentially, it's made up of two long oak crossties for the seating with two smaller pieces of the same timber bolted underneath for the legs. The bolts are countersunk and the holes filled with little oak plugs. I've planed the seating area smooth, and I've carved a ripple of waves on the surface along the front and sides. I find myself running my hands up and down the texture of the timber when I'm sitting there.

CONSIDERATIONS

- When selecting timber, it would be good to hand select as quality can vary. Look for timber that's not carrying too many large cracks or knots. It will, of course crack and twist with time, but it makes sense to start from the best place possible.

- I have used oak and left the wood in its natural state. The color will gradually turn a soft silver.

- The idea with the chiselled surfaces is that when you're sitting down, there is something very textural that you can run your fingers over.

- If you buy a crosstie at 7ft 10in or 8ft 6in (2.4m or 2.6m) long, you'll be able to cut the legs from the same piece of timber.

EXPERT INSIGHT

- I've used ¼in (6mm) spacers between the two crossties for the seat, but you could use ⅛in (3mm) spacers. The idea is to give them enough room to move as the wood ages, without rubbing against each other.

- By carving the front and back slightly differently, you'll get a different textural experience when you touch them.

- You'll be able to see shadow details on the carving at the front, so think about where best to position the bench to catch the light as it moves around the garden.

1 Measure and mark the dimensions of your four pieces of timber using a set square or carpenter's square, tape measure, and pencil.

2 It makes sense to mark your timber on all four sides. This will help stop you drifting as you cut the timber.

Cut to size

This bench is relatively cheap to make. Bear in mind that you can easily make it bigger or smaller to suit the space in your garden.

3 Use a circular saw or sharp handsaw to cut your timber. Make sure you follow your guidelines and take your time as you really need a good square cut.

4 If the surfaces are a little rough, you can use a hand planer on all four sides so it is smooth to touch.

6 Plane the edges of the timber on all the visible corners, working at 45 degrees to the edge. I look to take off about ⅛in (3 or 4mm). If you're not comfortable with the planer, you could just sand the edges if you prefer.

5 Next, use a belt sander to finish the timber surface. Alternatively, you could sand by hand using a block of wood wrapped in sandpaper.

7 Once you have finished all four pieces of timber, they should be smooth to touch. You can see here that the finishing brings out the detail in the timber.

> " *I was inspired by the idea of rhythm and waves for the carved textured surface on the front and back, but it could be whatever you fancy.* "

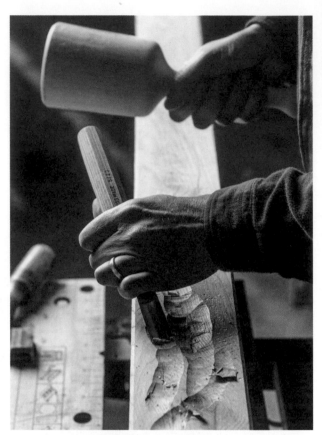

8 Use a pencil to mark the pattern you want to have on the front and back sides of the bench. I've used wavy lines to create a ripple effect.

9 Using the larger gouge chisel and carver's mallet, roughly cut out the larger sections of your pattern. Work along the grain of the timber. It may take a little working out, but this is a good time to get a feel for things as you can tidy up later.

10 Vary the size of your chisels to create waves of different depths. When carving, I tend to gouge out with the larger chisels to start with and then use the smaller chisels for details.

11 Once you're happy with your carving, you can take off any rough edges with a spokeshave. Use sandpaper to finish the finer details.

13 Create an even gap along the length of the two seat timbers by pushing in a spacer at either end. I've made a gap of ¼in (6mm) but you could make it a bit narrower or wider. The main thing is to allow room for the movement of the wood as it ages.

12 Now you can start to put the whole thing together. Place the two legs on the floor on their side. Put your two pieces of timber for the seat on top so they overhang the legs by about 4in (100mm) on either side. Adjust the position of the seat using a piece of 4in (100mm) timber as a guide. Use your mallet to gently knock the seat timbers into position.

14 Now mark a pencil line across the top of your seat at both ends, 6in (150mm) from each end. The line should align centrally with the leg below. Then measure 1½in (40mm) in from each side of the timber. These are your drilling points. Use a ½in (12mm) core drill to drill the holes about 1¼in (30mm) deep.

15 Then, using a ¼in (6mm) drill bit, make a pilot hole through the center of each ½in (12mm) hole. Firmly screw the seat to the legs beneath.

16 Put some wood glue on a block of wood, then collect a little on a spare screw and work it around the inside of the hole. This is to help hold the oak core plug in place.

17 Use a wooden mallet to gently tap the ½in (12mm) oak core plugs into the holes, leaving the plugs proud.

See life differently

Sitting on a low seat such as this changes your sightline, allowing you to see the garden and plants from a slightly different perspective.

19 Use a wood chisel—preferably a curved one—and wooden mallet to give the top of the oak core plugs a little texture.

18 Using a piece of cardboard to protect the surface of the bench, carefully saw off the tops of the plugs, again leaving them very slightly proud.

20 Finally, rub a small piece of sandpaper over the tops of the oak plugs to give them a smooth finish.

TOOL CHEST

This handy tool chest is basically a simple rectangular box with a lid that lifts off. It's about 55in (1.4m) long, 24in (600mm) wide, and 20in (500mm) high, and the lovely thing is it also doubles up as seating. I've painted the exterior in linseed paint—see page 143 for details on how to use this paint.

YOU WILL NEED

See p156 for timber requirements and measurements

- 2 and 2¾in (50 and 70mm) zinc or brass fixing screws
- linseed paint (I've used Brouns & Co. Maastricht Blue)

Tools

- workbench, vice, or clamps
- tape measure
- set square
- pencil
- saw
- cordless drill/screwdriver
- ⅛in (4mm) drill bits
- hand planer (if wood is very rough)
- belt sander or sandpaper
- paintbrush

I have several of these tool chests around the garden, and they are a great way of tidying away all sorts of stuff, such as outdoor cushions, barbecue equipment, kids' toys, and even beer! They are particularly useful in small spaces, which can often become cluttered. As well as providing extra places to perch, they are little understated focal points too. Inside the box I've fixed a couple of battens about halfway down to support a small shelf, which can be moved around to maximize space. Of course, you could buy something similar, but if you spend a weekend making this, you'll have a much greater feeling of satisfaction. The bench will last for years and you can customize it to suit the look and feel of your own space.

CONSIDERATIONS

- Remember the measurements are just guidelines. You can easily adapt this to fit your own space.

- Depending on what you want to use this tool chest for, you could easily add a floor to it. Just fix a second row of shelf supports ⅝in (15mm) up from the bottom of the box. Turn the box upside down and, using the same timber as the shelves, fix supports in place using 1½in (40mm) screws.

- The tool chest is really simple to maintain. Just give it a good clean once a year and repaint with linseed paint after a few years when it starts looking a little tired.

EXPERT INSIGHT

- It is a great place to tuck away everyday things such as hoses and clothes lines when you're entertaining and want the garden to look smarter.

- If you have a large garden, it's a perfect way to store tools that you use regularly but don't want to keep carrying around.

TOOL CHEST TIMBER

Use the patterns below to measure, mark, and cut all the timber so that you can assemble the tool chest much as you would a kit. You will need 1¾in (45mm) thick pressure-treated timber, 1¾in (45mm) thick scaffold boards, and roofing lath.

FRONT VIEW

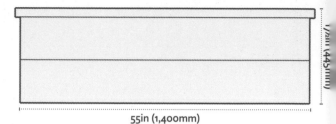

1½in (445mm)

55in (1,400mm)

SHELF SUPPORTS x 2
1¾in (45mm) thick roofing lath (batten)

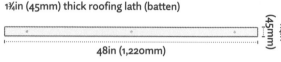

1¾in (45mm)

48in (1,220mm)

SIDE PANELS x 4
1¾in (45mm) thick pressure-treated timber

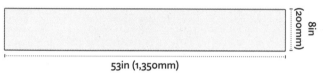

8in (200mm)

53in (1,350mm)

SHELF PANELS (OPTIONAL)
1¾in (45mm) thick pressure-treated timber

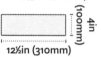

4in (100mm)

12½in (310mm)

LID PANELS x 2
1¾in (45mm) thick scaffold boards

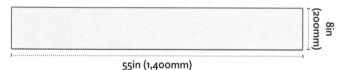

8in (200mm)

55in (1,400mm)

FIXING BLOCKS x 4
1¾in (45mm) thick pressure-treated timber

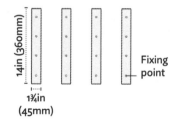

14in (360mm)

1¾in (45mm)

Fixing point

END PANELS x 4
1¾in (45mm) thick pressure-treated timber

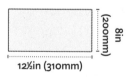

8in (200mm)

12½in (310mm)

LID BRACKETS x 4
1¾in (45mm) thick roofing lath

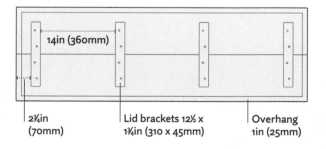

14in (360mm)

2¾in (70mm)

Lid brackets 12½ x 1¾in (310 x 45mm)

Overhang 1in (25mm)

1 Following the timber list on the opposite page, use the tape measure, set square, and pencil to measure and mark the timber to the required lengths.

2 Then, using a saw, cut the wood to length. I prefer to cut all the pieces at the same time so that I can put the whole thing together like a kit.

3 Lay two side panels on your workbench with the best side facing down. Make sure the ends are aligned. Next, offer up one of the end panels and sit it flush with the side panel. Now, place a fixing block next to the end panel, making sure the block is flush with the side panel nearest you. Finally, mark the position of the fixing block with a pencil line.

4 Then remove your end panel and realign the fixing block with your pencil mark. Next, use four 2¾in (70mm) screws to fix it in place. Repeat this process at the other end of the side panel. Then repeat steps 3 and 4 for the other two side panels.

Built-in seating

You could use the same principles and skills used here to make built-in seating. I have also made this on a slightly bigger scale to use as covered sandpits for the kids.

5 Put one of your side panels on your workbench with the fixing blocks facing up. Then line up a long batten to support the shelf just below the halfway point. You can, of course, adjust this depending on what you want to store. Using 2in (50mm) screws, fix the batten about every 12in (300mm), starting about 2in (50mm) from the end. Repeat this step for the other side panel.

6 Lay the two scaffold boards for the lid on your workbench with the best sides facing down. Make sure the ends are flush. Mark the center of each of the four lid brackets with a pencil. Starting 2¾in (70mm) in from either end, position the brackets evenly along the lid, making sure the bracket's center and lid's center align. Use a pencil to mark each bracket's fixing points.

7 Using your pencil marks as a guide, first fix the two end brackets to the lid with 50mm (2in) screws. Then fix the other two brackets in place. If you struggle to hold the timber firmly, you could use clamps as a temporary support. The brackets will sit snugly inside your box and stop the lid moving.

8 Stand the two side panels upright and insert the end panels one at a time, making sure the cut ends sit inside the side panels and butt up to the fixing blocks. Fix the bottom piece at each end first using 2¾in (70mm) screws. This helps stabilize things before you fix the upper two ends. If possible, have someone hold the pieces in place while you are fixing.

9 Now offer up the lid just to make sure everything fits snugly.

11 Paint the outside of the tool store with three coats of linseed paint. Then, once everything is dried, you can put your tool store in position.

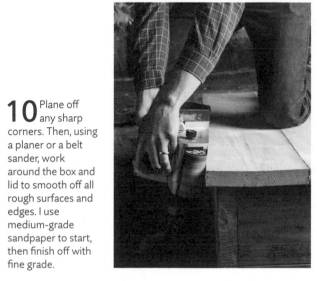

10 Plane off any sharp corners. Then, using a planer or a belt sander, work around the box and lid to smooth off all rough surfaces and edges. I use medium-grade sandpaper to start, then finish off with fine grade.

12 Finally, insert your shelving. Remember that the pieces of wood are not fixed so the position can be adjusted. You can add more shelf space if you wish.

PROJECTS

WATER FEATURES

> *Within a few weeks of finishing the pond, a frog moved in, and I didn't even know we had frogs in the garden.*

INTRODUCTION

Almost the moment you add water to your garden, it attracts life: birds, insects, amphibians, and mammals, including humans—we all love it. What's fantastic about water is its versatility. It can be still and calm with the occasional ripple, or it can move fast in jets or sprays and animate the space with movement and sound.

Water's reflective qualities are magic too, and, like fire, it has the power to draw people toward it. When you introduce water into your garden, it brings another layer of interest that helps make the whole place extra special. My project for Mrs. Frost's vintage pot is a great little starting point. It's very easy to achieve, and you can either use it by itself or in multiples to create little pause points around the garden. The round pool is a bigger investment in terms of both time and money, so it's important to take the time to decide where best to put it and what you want from it. I wanted a tranquil pool in an area of the garden we generally use later in the day, and interestingly, now when I sit there, everything starts to seem that little bit simpler and calmer.

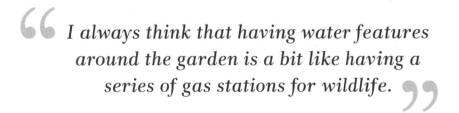

I always think that having water features around the garden is a bit like having a series of gas stations for wildlife.

MRS. FROST'S VINTAGE POT

Even the smallest amount of water in a garden brings it to life. Containers such as this galvanized tub filled with aquatic plants work like miniature ponds and are ideal for attracting wildlife to the garden. They are particularly good if you have limited space, or you could use several versions in a bigger garden.

YOU WILL NEED

- galvanized container 16in (400mm) deep and 18in (450mm) in diameter
- aquatic pots
- aquatic potting mix
- aquatic plants and oxygenators— I've used two aquatic plants and one oxygenator here (see below)
- gravel
- stones or bricks for steps
- 1¼–2in (30–50mm) cobbles or pebbles

Tools
- watering can or hose
- trug

Plants I used in this project
- pygmy water lily (*Nymphaea tetragona* 'Alba'); smooth iris (*Iris laevigata*); hornwort (*Ceratophyllum demersum*)

I think most of us have an innate fascination with water. As well as inviting in wildlife, water can bring sound, movement, and calm. If you put them in the right place, water features can become beautiful reflective surfaces and are a great way of slowing people down as they wander around the garden. I tend to use them as pause points where people can sit. Think carefully about where best to put your water feature so it doesn't overheat and so that the plants and any creatures who use it as habitat will be happy too. It's amazing how much wildlife even a small water feature can attract.

CONSIDERATIONS

- This container is an old galvanized steel boiling pot that belonged to Mrs. Frost, who loves things like this. I've also made this water feature with baths, tanks, and old rain barrels. Just make sure your container is watertight and an adequate size.

- I probably wouldn't go much smaller than the container I've used here as smaller volumes of water tend to heat up too quickly. If I used a more shallow pot, then I would make sure it was wider and put it in a shaded spot, or plant more of the surface to cover it. The quantity and depth of water as well as the location will affect the amount of evaporation, but you can often manipulate things to get the best result.

EXPERT *INSIGHT*

- **Check the label of aquatic plants or do some research to find out how deep the top of the root ball should sit below the water. They all vary slightly—some plants are fully submersible, others semi-submersible—which is why it's useful to create little shelves within the container that allow you to set the depths accordingly.**

- **Give the plants time to settle down, and watch how things evolve. You might find that you need to add another plant or more oxygenators, but usually things find their own balance.**

1 First, make sure your plants are in decent-sized aquatic pots. Here I'm using a half-gallon (2-liter) pot for an iris I split from a clump. Put a few handfuls of aquatic potting mix at the bottom so the plant has something to sit on.

3 Top it off with a layer of fine gravel. This helps hold the potting mix in place and stops it floating away when you put the pot in the water.

2 Fill any gaps around the sides with more aquatic potting mix and make sure there is good contact with the roots. Give the pot a little tap to settle things in.

4 Fill the container about one-third full with water. Then put a row of stone blocks about half or two-thirds of the way around. The idea is to create a little staircase so that wildlife can make their way in and out of the water.

5 Start the next row a bit further round, which forms the first step. Then keep repeating rows of staggered blocks until they reach about 4in (100mm) from the top.

6 The stone blocks also provide lots of little shelves to put the plants on and allow you to adjust the heights as needed. Put your plants into position according to how deep they prefer to be submerged.

7 Use a final stone block on top of the edge of the pot or pots to hold them in place. This also helps to provide access for wildlife.

8 Empty your bag of pebbles or cobbles into a trug and wash off any dust and debris. It's important to get everything you're using in the water feature as clean as possible.

11 Next, I'll put in a pygmy water lily. Water lilies need to be planted in stages so that you don't drown them. If you dunk them straight in and all the leaves go under, the plants will suffer. So, for the short term, I've placed another brick in the bottom of the container on which to put my water lily.

9 Then put cobbles in to bury the staircase and add a bit of detail and dressing to the tops of the plants. Cobbles also give creatures places to perch and hide in the water.

10 Fill up the container, ideally with rainwater. If you don't have rainwater, you can just use water from the tap. I tend to let it overflow so that any surface debris runs off and leaves everything clean.

12 Initially, the water lily sits just below the surface. Over the next few days, I'll lower it into its final position.

14 Give it time to settle down and find its own balance. If the water turns green, leave it for a while as it may clear as the plants grow. If not, you may need to add more oxygenating plants or move the feature.

13 Next, put in the oxygenating plants. These help to keep the water clean and add vital oxygen. It will be the balance of planting that keeps your water clear, and it may take a little adjusting and time to settle down.

Handy handles

There might be some trial and error to get the container in the right place. The benefit of having handles is that you can empty some of the water and carry it to a new spot in the garden.

15 Think about where you put your water feature. Do you want to use it as a reflective surface, as a little pause point, or put it somewhere out of the way to encourage wildlife to visit? Make sure that the container won't get too hot and heat the water up, which will cause the water to evaporate or allow algae to grow.

ROUND POOL

A pool of water in the garden creates a great focal point, is a lovely reflective surface, and draws wildlife into the garden. Inspired by the circles of mushrooms called fairy rings that appear like magic in nature, this simple circular pond has sloping sides and stone detailing around the top.

YOU WILL NEED

Use the instructions on pp244–45 to calculate quantities for:

- concrete (1:10 cement to ballast)
- soft sand to cover the pool with a 2in (50mm) layer
- fleece liner
- rubber liner
- 2¾ x 4in (70 x 100mm) stone for edging
- light-colored mortar (1:5 cement to soft sand)
- 1¼–4in (30–100mm) pebbles

Tools

- line and pins or spray line
- string
- wheelbarrow
- spade and shovel
- pick axe (for digging hard ground)
- wooden pegs
- hammer
- level
- plasterer's trowel and small trowel
- tamping board
- hose
- plank with screw or bolt in the center (to lay across pond)
- tape measure
- sharp knife
- fork

This circular pool is about 10ft (3m) in diameter and 28in (700mm) deep, although you could make it pretty much whatever size you fancy. I excavated it so that the sides are sloping and form a shape a bit like an upside-down derby hat. It's lined with a sheet of rubber to make it watertight and this is held in place by a circle of stone around the perimeter. To make sure that wildlife can get in and out easily, I've added a cobble bank and a little perimeter shelf.

CONSIDERATIONS

- The best location for a pond is in a clear, open space that gets a reasonable amount of sunshine but won't overheat. If you do have to put your pond in a place where the water might overheat, cover the surface or make the pool deeper.

- The soil you dig out to create the pond can either be spread around the garden or you can use it to fill raised beds. Alternatively, you might want to dispose of it offsite.

- The stone edge defines the pond's perimeter and makes a neat edge for a lawn. It also picks up on other designed elements in the garden. Make sure your mortar mix sits comfortably with your stonework. Here, I've used white cement and a light sand to achieve a pale finish.

EXPERT *INSIGHT*

- Before excavating the site, it's worth digging a little test pit to gauge its suitability. If you're trying to dig out solid rock, for instance, it's best to choose another location.

- Always put in some oxygenating plants to keep the water clear. I use the shelf to plant a few verticals as they help to break the line formed by the edge.

- If you want to introduce some movement and sound, rather than a still, reflective pond, you could install a small pump or fountain.

CROSS-SECTION OF THE ROUND POOL

This simple cross-section gives guidance on how to dig and shape your pond, and the different layers of sand, fleece, and liner you need to add.

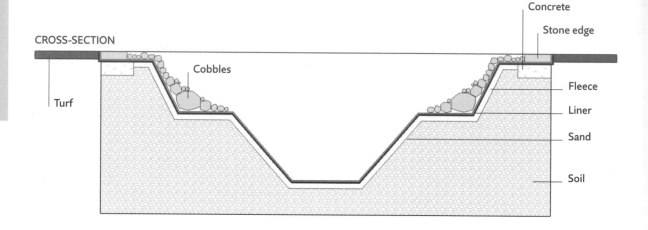

CROSS-SECTION

Concrete

Stone edge

Cobbles

Fleece

Liner

Sand

Soil

Turf

2 Knock in pegs outside the pond area to check the level of the ground around it. This will help you understand your site and where you may need to make up levels. They can also guide your finished level. Use a level to make sure the pegs are all the same height.

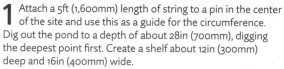

1 Attach a 5ft (1,600mm) length of string to a pin in the center of the site and use this as a guide for the circumference. Dig out the pond to a depth of about 28in (700mm), digging the deepest point first. Create a shelf about 12in (300mm) deep and 16in (400mm) wide.

3 Use string and your central pin to measure about 4ft (1,200mm) from the center. Use spray line to mark the circumference of your concrete footing.

4 Next, tap in a series of pegs just outside the spray line to set a level for the concrete footing. I have set the pegs 5in (130mm) lower than my finished level, which allows for the stone, mortar, liner, and protective fleece.

5 Next, you need to put a concrete footing in all the way around the top of the pond to provide a flat and solid surface for the stone edge. Mix up the concrete so that it is reasonably wet and then shovel it into place to a depth of around 3–4in (80–100mm). It's important to tamp it level as you go before it sets.

Check it out

Always check measurements and levels for water features, and then check and check again. Water will make any mistakes in level very clear, and it's impossible to undo these mistakes or fudge them.

7 Tamp it down with a piece of wood, using the pegs as a guide for the level all the way around. Leave it to set for at least 24 hours.

6 Then use a plasterer's trowel to shape the concrete around the pool's edge.

8 Create your shelf for the cobbles, adding extra soil if you need to. My shelf is about 12in (300mm) below the concrete level and about 16in (400mm) wide. Firm the soil in the bowl back into shape, filling in any holes you may have left.

9 Line the pond with soft sand, using a plasterer's trowel to make sure the surface is smooth and even. This will provide extra protection for your liner.

11 Next, carefully position the rubber liner in the pond. To begin with, the liner should come up a fair amount over the footing. Once you have everything in place, you can then start to fill the pond. You may need to adjust the liner in places.

10 Carefully put the fleece liner in the pond so that it drapes over the top edge.

> *Adding water to the garden helps bring it alive by providing habitat for wildlife and creating atmosphere.*

13 Attach a screw or bolt into the center of a long plank of wood and tie on a tape measure. Position the plank so that it runs through the center of the pool so that you can check the measurements with the tape. This will guide the outside of your stonework. You may have to adjust this a little to get it right.

12 If you have to make folds in the liner to make it fit, try to make them in a concertina pattern rather than big single folds. Fill the pond about three-quarters full to pull everything down. Adjust the folds to make them as even as possible.

14 Make sure you lay the mortar on a day when no rain is forecast. Use the tape measure like the hand of a clock so that it acts as a guide for the circumference of the pond. Lay down a bed of mortar on top of the liner and start to set your stones in place, keeping the pieces nice and snug. Don't worry about the joints at this point, as you'll sort them out later. Here, I've used the stones on their edge to create a tight circle.

15 Continue laying the stones, making sure you keep checking the circumference against your tape measure.

16 Use a short level to check the stones are level in both directions.

17 Use a long level to check all the stones are at the right height. This is really important, so take your time. Once you have everything laid and checked, you can start your pointing.

18 Use a small trowel to point the joints between the stones with mortar, making sure you fill the gaps evenly and neatly. Here, I've used a light-colored soft sand mixed with white cement so that the pointing sits comfortably with the stonework.

19 Leave the mortar for 24 hours until it has set solid. Then fold the liner up the back of the stones. Use a knife to cut away and tidy up any excess liner, being careful not to cut too low.

21 Check that any cement or sand hasn't fallen in the pond. If need be, empty the pond to remove it and give it a good wash-out. Clean your cobbles thoroughly in a bucket of water. Then build your cobble beach, which will provide access and habitat for wildlife, and also help to hold your plants in place.

22 Once all your cobbles are in place, you may want to give your pond one last clean before you start planting. If not, you can continue to fill, ideally with rainwater but tap water will also be fine. Once you've filled your pond, you can start to add your plants (see box opposite). The pool may go green after a while, but give it time, and the water will start to clear when your balance of planting is right.

20 Clean any excess mortar off your liner. Then backfill soil around the edge so that the stones are at ground level. The soil also helps hide the edge of the liner and holds it in place.

WATER PLANT IDEAS

When choosing plants, consider those that not only provide flower interest but also habitat, oxygen, pollen, and nectar throughout the season. Here are a few of the plants I've added to my pool. It's a simple mix that you can personalize to your taste. When you're planting marginal plants, check the label to find out whether the plant can go under the water.

- **Water mint** (*Mentha aquatica*; below left)
  A great marginal perennial, if a little vigorous, it offers pollen for wildlife when in flower. It's happy in sun or shade and grows to about 20in (50cm) tall.

- **Marsh marigold** (*Caltha palustris*; below center)
  This herbaceous native perennial grows to about 20in (50cm) tall. Lovely round green leaves sit below rich yellow buttercup-shaped flowers. It flowers in late spring and is happy in a sunny spot.

- **Dwarf cattail** (*Typha minima*)
  I do love this plant—it's a miniature version of the much-loved bulrush, making it ideal for smaller pools. It has slender leaves and grows to about 24in (60cm) tall. The distinctive brown spikes appear in July and August.

- **Yellow flag** (*Iris pseudacorus*)
  This iris is a shallow-water plant. It's a strong grower—reaching a height of up to 5ft (1.5m)—and clumps well but is easy to contain in a water basket; plant it in a container that's 8in (20cm) across. It has swordlike leaves and strong yellow flowers in summer followed by brown seedheads.

- **Marsh cinquefoil** (*Potentilla palustris*)
  Marsh cinquefoil spreads across the surface so it's great for creating cover for wildlife and shade for the water. Its leaves resemble those of strawberry plants and it's sometimes called bog strawberry. Clusters of dark red flowers appear in summer. It grows up to about 20in (50cm) tall.

- **Creeping spearwort** (*Ranunculus flammula*)
  This is a hardworking marginal plant with cracking yellow cup-shaped flowers from May to August; it grows to about 3ft (1m). Its creeping habit makes it a winner with wildlife as it provides cover. Be careful when handling as it can irritate the skin.

- **Hornwort** (*Ceratophyllum demersum*; below right)
  Every pool will need oxygenating plants and hornwort tends to be my starting point. It has dark olive-green needlelike foliage. A good coverage of this plant helps keep algae down. When it gets to about 30 percent coverage in the pool, I just take some out. It can grow up to 4in (10cm) tall.

Water mint (*Mentha aquatica*)

Marsh marigold (*Caltha palustris*)

Hornwort (*Ceratophyllum demersum*)

PROJECTS

HARD LANDSCAPING AND DESIGN FLOURISHES

> *Add detail, flourishes, and color to your garden so it feels uniquely yours.*

INTRODUCTION

The hard landscaping in our gardens is sometimes treated like the poor relation, but actually, I think it really can set the tone and feel for the garden. It does give you a great opportunity to personalize your space. For me, all the best gardens have a decent structure that feels comfortable in terms of the materials and proportions.

Hard landscaping elements form what are often called the "bones" of the garden, and they can help establish the look and feel of the space before any plants are added. But they can be beautiful as well as practical and, if you personalize them, they are a great way of giving your garden a unique character. It doesn't take much— even small details can make a big difference. For instance, you could lay the bricks on your terrace or paths in a herringbone pattern, chisel a design into the face of your crosstie steps, or add posts and wires to the corners of a raised bed that will support fleece or mesh. Think about color, too, and how you can tie it in to other elements in the garden with paint or by charring the surface of wood. All these little flourishes add a distinct character that makes the garden sing.

" Homemade things in your garden not only make it more characterful but also provide lovely memories of the time spent working on them. "

WOODEN SCREEN

These wooden screens are made from sawn pressure-treated timber boards, but you could use other types of wood, scaffold boards, or even pallets. Inspired by designer Paul Smith's signature stripe, I cut the boards into different widths. The narrowest is 4in (100mm) wide and the widest is 8in (200mm).

YOU WILL NEED

- sawn pressure-treated timber boards roughly 1in (25mm) thick and 12⅓ft (3,750mm) long cut to different widths:
  5 boards 8in (200mm) wide
  4 boards 6in (150mm) wide
  2 boards 4in (100mm) wide
- sawn pressure-treated timber posts 4 x 4in x 8½ft (100 x 100 x 2,600mm) used every 6ft (1,800mm)
- concrete mix (10:1 ballast to cement) or Postcrete
- 1½in (40mm) screws
- extra timber for temporary props
- weatherproof paint

Tools

- tape measure
- line and pins or spray line
- posthole digger
- spade or shovel
- level
- straightedge
- mixing tray or concrete mixer (optional)
- timber battens
- cordless drill/screwdriver
- ⅛in (4mm) drill bit
- tile spacers
- sandpaper
- paintbrush

The screens can be used in various different ways, such as for fencing, screening something unsightly, or creating dividers between areas. Having the boards in different sizes makes the screen more characterful and unique. And by attaching the boards to the uprights horizontally (rather than vertically), you don't have to use cross rails between posts, which adds strength and makes it far simpler to build. You can leave as much or as little space between the boards as you like, depending on the effect you're after. I've painted these screens, but you could leave the wood to silver naturally, stain them, or try out the Japanese technique of charring the wood called *shou sugi ban* (also known as *yakisugi*).

CONSIDERATIONS

- I've installed these boards horizontally and positioned the uprights every 6ft (1.8m), but there are lots of different styles of fixing them together that you could play around with. For example, they could be attached vertically or at an angle. You could also leave spaces between each board. These are around 6ft (1.8m) tall, but you could make them just 3ft (1m) tall or have a mix of different heights.

- If you want to grow plants up the panels, screw in some eyes and fix horizontal wires on to them so plants can be supported discreetly.

EXPERT *INSIGHT*

- **Anything that creates a garden boundary has a huge effect on the atmosphere and style of the garden. The more you can make your fences "disappear" by using color and planting, the better your garden will look.**

- **Bear in mind that lighter colors appear to come toward us, while darker colors seem to disappear and are particularly good as a backdrop to plant against.**

1 Measure the site where you will put the panels, clear it of any weeds, and make sure it's level. Set up a tight string line, making sure the end pins sit past the point where you will add your end posts. If you need to make sure it's square to the house, use the 3-4-5 triangle method (see p240).

2 Spray lines or use sand to mark the positions for the posts—I've put mine 6ft (1,800mm) apart.

3 Use a posthole digger or spade to dig out the holes. Keep the hole relatively tight around your post. I've made mine 12 x 12in (300 x 300mm) wide and 28in (700mm) deep. As I'm using heavy boards, I dropped the posts in slightly deeper than I would normally.

Putting in posts

Don't dig all of the holes for your uprights in one go. It's really easy to get your measurements slightly out and then you end up doing double the digging. It's far better to do just two or three at a time.

> *Anything that creates a boundary in a garden has a huge effect on the whole atmosphere and overall style of the garden.*

5 Mix the concrete (or use Postcrete). Fill the hole with concrete around the post, tamping it down with a piece of spare timber to ensure there are no air pockets. Use a level to make sure the post remains vertical and is sitting next to your string line.

4 Put your post in the hole and make sure it's sitting vertically using a level and your setting-out line. You can pencil mark your post 6ft (1,800mm) from the top to help guide the height of the posts. It's also useful to use a straightedge along the top of the posts to make sure they are level with each other.

6 Then screw battens in place to hold the post firmly in position while the concrete sets. Repeat the process for the remaining posts, making sure they are vertical and level across the top with each other. Leave the concrete to set for 12–48 hours.

7 Next, fix the first board between the posts, making sure everything is straight with a level. I've set mine about 2in (50mm) above ground level and used blocks of stone to support it.

8 Attach the boards with screws (two screws per post). I draw a vertical line to guide the postion of the screw holes. Use a tile spacer to give a small, even gap between each board. It adds a lovely shadow line and allows the timber to expand and retract.

9 Continue screwing the boards in place, and keep checking everything is level and even with your level. You may need to adjust a little as you go along. I have let the board run past the upright post at one end in order to hide the post. On a corner, always make sure the cut edge is hidden and the best-looking edge faces outward.

10 If you wish, you can add another panel at right angles to the first one to create a corner screen. I like the screws to run in a vertical line for a neat and tidy finish.

11 Check over the surface and smooth off any rough edges with sandpaper. Next, using a wide brush, apply two coats of paint. The color softens the screen in its landscape. You could also try the charring technique, depending on the finish you want.

12 Stand back and check over your handiwork. You may need to do a final touch-up of paint so the screen looks as neat as possible.

Charring technique

Instead of painting, you can lightly burn the timber so it turns brown or do it more intensely until it starts cracking. You can apply the technique to all sorts of items—see pp77 and 87, for instance.

FIRE PIT

I love fire pits and have made many of them over the years. They take me back to being a kid and the simple pleasures of just sitting round a campfire with my mates. Maybe there's something primeval about it—fires draw people toward them and change the atmosphere of a whole space.

YOU WILL NEED

- stones or old brick
- wooden pegs
- plastic pipe or pot, about 6in (150mm) in diameter
- gravel
- concrete (1:8 cement to ballast) or Postcrete
- small to medium-sized limestone blocks, or other rock that can withstand heat
- mortar mix (1:4 white cement to soft sand)
- plasticizer

Tools

- tape measure
- set square
- spray line
- line and pins
- spade
- hand sledge
- level
- knife
- concrete mixer or mixing tray
- watering can or hose
- small tamping board or float
- safety goggles or glasses
- masonry chisel
- brick trowel
- wire brush

Fire pits are a great way to extend the time you can spend in the garden, particularly when the nights are drawing in or it's a bit chilly. Just make sure that you put your fire pit in a sheltered spot that's comfortable to sit in, and you'll use it often. This one is really simple. It's essentially a little square of bricks measuring about 32 x 32 x 12in (800 x 800 x 300mm) and built so the top sits at ground level. It could be made out of a different type of brick or as a stone wall. I used limestone because it's local to my area and I have plenty in my garden. Try to choose materials that reference something that's in the garden to give it cohesion.

CONSIDERATIONS

- You want the fire pit to be beautiful as well as practical. It can be pretty much as big or as small as you like, but the main thing is to think about how you plan to use it—whether it's for toasting marshmallows, as a barbecue or a spit roast, or just to create a warming fire. There's loads of possibilities, but if you want to cook on it, then it's worth researching what size grills and other hardware are available.

- If possible, position the fire pit so that it benefits from a good view—sitting in front of the fire watching the sunset is magical, for instance. On the edge of woodland, at the bottom of the garden, or a gravel garden also make good settings.

EXPERT INSIGHT

- The fire pit doesn't have to be square—it could be round or rectangular—but the shape needs to work with your garden design.

- This fire pit measures 32 x 32in (800 x 800mm) so there's plenty of room for good air circulation to keep the fire alight. I've put a pipe in the concrete base for drainage.

OVERHEAD AND CROSS-SECTION OF THE FIRE PIT

This square fire pit sits at ground level with a drainage pipe at the center. Here, I've used a bed of concrete for the base and three courses of stone including a capping layer that's 4in (100mm) high and 4¾in (120mm) across.

OVERHEAD

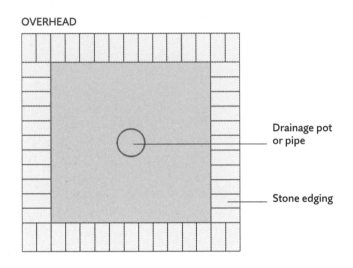

Drainage pot or pipe

Stone edging

CROSS-SECTION

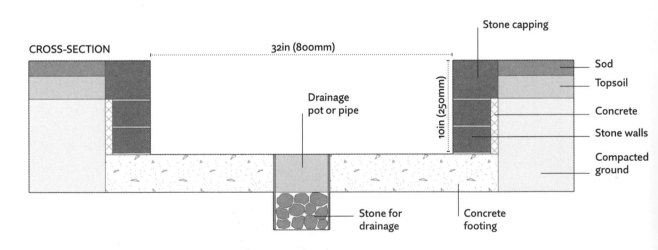

Stone capping

32in (800mm)

10in (250mm)

Sod

Topsoil

Concrete

Stone walls

Compacted ground

Drainage pot or pipe

Stone for drainage

Concrete footing

1 Choose a location for the fire pit that's sheltered from the wind. Measure out the site using a tape measure, set square, spray line, and line and pins.

2 Dig the marked-out area to a depth of about 12–18in (300–450mm). Always make sure that you dig out more than you need—4in (100mm) all the way round should be fine. Then tread all over the area to tamp the base down so it's firm.

3 In the center of your pit, dig another hole another 16in (400mm) deep with a diameter of about 12in (300mm). This is for the drainage pipe or pot.

4 Part-fill the hole with stone or old brick, leaving enough room for gravel and the plastic pipe or pot.

" Having a fire pit means you can stay outdoors much longer, particularly as the nights begin drawing in and it's a bit chilly. The main thing is to put it somewhere sheltered and comfortable. "

5 Knock in wooden pegs to set the level for the concrete across the bottom of the hole. Allow for 4in (100mm) of concrete, leaving roughly 10in (250mm) including the mortar joints to the finished wall.

6 Use a tape measure and level to check that the pegs are straight and level across the area.

7 If using a pot, remove the bottom with a knife, then put the pot in the center of the hole and shovel in some gravel to hold it in place. Use a level to make sure the top of the pot sits level with the top of the timber pegs.

8 Mix the concrete in a concrete mixer or wheelbarrow, gradually adding water until it is a sloppy but not runny consistency. It is important that it is not too dry or you will find it difficult to level. Shovel the concrete into the hole, taking care to spread it out across the dug out area.

⚠ **SAFETY FIRST**
Wear safety goggles or glasses when cutting stone with a masonry chisel.

10 If you need to cut blocks to size, measure and mark up on all four sides with a pencil and set square. Then align the masonry chisel along the pencil line and give it a sharp blow with the hammer. Turn and repeat on each side. This will split the stone evenly.

9 Once the concrete is roughly level, use a strong piece of board, slightly smaller than your base, to tamp it level, working to the top of the timber pegs. Let it set for 24 hours.

11 Set up lines on all four sides, using the 3-4-5 triangle method (see p240) on each corner. Take your time with the stonework. Lay your first course on a mortar bed of 1:4 white cement and soft sand with added plasticizer to make the mix more pliable. Work your corner up first, making sure the stones are both vertically and horizontally level.

13 As you lay the second course, keep checking the vertical as well as the horizontal levels.

12 Point joints with the mortar mix as you go along. I tend to build the corners first, making sure they are spot on before completing the sides.

14 Once you've laid two courses of stone, you need to put on a capping layer with pieces of cut stone laid on edge. This is to protect the stone below and provide a neat edge to the fire pit. Use a hammer and masonry chisel to cut the stone. Mine are 4in (100mm) high by 4¾in (120mm) across.

16 Point when the mortar mix is damp. I use an old piece of hose and a small gouging trowel. Once all the pointing is complete, clean up the stonework with a wire brush. Brush the surfaces clean and leave the whole thing to set for 24 hours before using.

15 When the capping is complete, cement the back of the stones to create a haunch, which will add further support. The haunch will be buried when you put the topsoil back.

17 For a neat finish, put topsoil back around the perimeter of the fire pit and lay it with sod.

CROSSTIE STEPS

I've seen a lot of crosstie steps in my time and wanted
to create something that was a bit more interesting.
The risers are made from crossties set on their side and
the tread is simply lawn sod. They are a great way of
dealing with level changes in more informal areas of the
garden that don't have a huge amount of footfall.

YOU WILL NEED

**Use the instructions on p200 and
pp244–45 to calculate quantities for:**

- 8 x 4in (200 x 100mm) crossties,
  cut to 24in (600mm) lengths
- 4 x 2in (100 x 50mm) pressure-
  treated timber posts, cut to about
  20in (500mm) tall (2 per step)
- 4in (100mm) timber locks or screws
  (2 per each support post)
- concrete (1:8 cement to ballast)
  or Postcrete
- topsoil and lawn sod

Tools

- line and pins
- measuring post taller than
  the height of your slope
- long level
- tape measure and pencil
- carpenter's square
- Handsaw
- gouging axe or 2in (50mm)
  gouge chisel
- blowtorch
- shovel
- bucket or mixing tray
- brick trowel
- hand sledge
- tamping board
- wood for measuring riser height
- cordless drill/screwdriver
- sharp knife
- rake

These steps are pretty understated so they are great for wilder areas of the garden. I've made them a little more eye-catching by putting wildflower sod either side and adding texture to the wood with a gouge axe to take out small chunks. I've also burned the surface to produce a lovely blackened finish, and the colors, which go from blacks to browns to the natural wood, look great against the green of the sod. I've used five steps here, but you'll need to measure and calculate how many you require depending on the length of your slope (see p200).

CONSIDERATIONS

- Blackening the wood helps to seal
  and protect the timber, creates an
  attractive effect, and is really easy to
  do. Use a small blowtorch and make
  sure you are in an open spot outside
  with plenty of space around you. It's
  a good idea to practice beforehand
  on a spare piece of wood. In Japan,
  I've seen timber burned deep into
  the wood so it's really black with
  cracks in it, but you could do it quite
  lightly so it just takes the shine off.

- The length of the support posts
  depends on your soil—if it's soft,
  they may need to be longer as you'll
  drive them quite far in; if it's hard,
  you might struggle to drive them
  in very far at all.

EXPERT *INSIGHT*

- **You should always design
  paths and steps with
  the main purpose of getting
  from A to B, but by making
  them narrower, you can
  influence how people move
  through the space, slowing
  them down and making the
  journey a little more engaging.**

- **When you are setting up any
  steps, measure, measure,
  and measure again.**

- **Instead of using sod, you
  could put in bark or gravel or
  lay pavers or bricks. It's easy
  to change the steps to meet
  your needs.**

CALCULATING STEPS

To calculate how many steps you need to comfortably ascend or descend a slope, first measure how far the steps will extend—the horizontal distance. You need to measure from fixed points at the top and bottom of your slope. Then measure the height at which your steps start and finish—the vertical distance. Then divide the vertical distance by the riser height to calculate the number of steps. For the riser height, aim for 6–8in (150–200mm). Next, divide the horizontal distance by the number of steps to find the tread depth. I work on a minimum tread depth of 12in (300mm). Whatever dimensions you choose, make sure that all your risers are equal in size so they aren't a trip hazard.

OVERHEAD

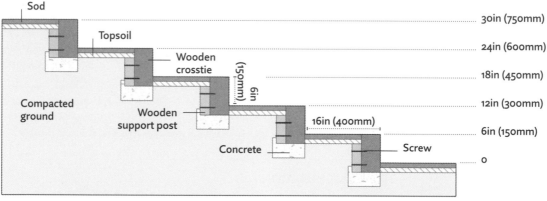

Sod

Topsoil

Wooden crosstie

Compacted ground

Wooden support post

Concrete

6in (150mm)

Screw

30in (750mm)

24in (600mm)

18in (450mm)

12in (300mm)

6in (150mm)

0

16in (400mm)

CROSS-SECTION

1 Prepare your site by putting in a line and pins to mark the position of your steps. Put the pins past the first and last step, otherwise you will have to move them when building. Run a line nice and taut between the pins.

2 Bang a post in at the bottom of the slope. You will use this to help you visualize, calculate, and if necessary adjust the height of the risers. It's really handy to refer to and check measurements as you build.

3 Put a level vertically at the front of where you plan to put your first step and use a tape measure to work out the horizontal distance between the top and bottom of the slope. This will enable you to work out the size and number of steps.

> *These steps were designed for a little-used place in the garden. If they were going to have more footfall, then I'd probably use bark for the treads instead.*

5 Use a carpenter's square and pencil to mark your risers and cut them with a sharp handsaw. I've cut mine at 24in (600mm) wide. Prepare the risers by giving the two sides that will be exposed (front and top) some texture and color. First, use a gouge axe to take out small chunks. You could also achieve this with a large gouge chisel. You can take out as much or as little as you like, but just make sure it's even.

4 Next, mark on your measuring post the vertical height of the slope. To work out the number of steps, divide the vertical height by the riser height. I usually aim for around 6in (150mm) per step and generally won't go over 8¾in (220mm). Mark the height of your risers and the number of steps on the measuring post. Next, divide the horizontal distance by the number of steps to find the tread depth. Treads should not be less than 12in (300mm) deep, but you can be more generous if you have the space.

7 Now dig out the position of your first crosstie. Dig out to the depth of your crosstie plus 2in (50mm) for the concrete.

6 Next, to give them a little color, use a blowtorch to blacken the surface. The degree to which you do this is really a matter of personal taste. It's a good idea to practice on a spare piece of the same wood first so that you get to understand how the wood responds to burning. Take your time and make sure you have plenty of space around you.

8 Then drop the crosstie into the hole and use a level horizontally from the post to measure the height of your first step. Double check that you have allowed 2in (50mm) for the depth of the concrete it will sit on.

9 Lift the crosstie out of the hole. Mix up your concrete or use Postcrete and lay a bed about 2¾in (70mm) thick. Work it roughly level with your trowel. This will give you enough depth so you can tamp your crosstie into place.

10 Cut your support posts from pieces of 2 x 4in (50 x 100mm) timber. I've made these 20in (500mm) long, but you may want to make them longer depending on how soft your soil is. The important thing is that you drive them into the ground until they are firm and solid.

11 Lay your first riser on top of the concrete and use a hammer to gently tamp it down until it sits at the correct height. Use a level to make sure it is level both horizontally and vertically. Also make sure it is square to the line.

13 Use your trowel to add in more concrete to hold the support posts in place.

12 While the concrete is still wet, hammer in two support posts deep enough so that they are firm and you can't rock them about. At this point, they should sit proud of the top of the step.

14 Use a piece of wood to tamp down the concrete and eliminate any air pockets. Then leave the concrete to set for 24 hours. Alternatively, rather than wait 24 hours for the concrete to set each time you put in a step, you can build all of the steps, securing them with support posts first, and then put concrete behind all of the steps at the same time.

> *If you try to drive in supporting posts flush with your sleeper, you can guarantee that you will take chunks out of the top of your sleeper. So keep them proud, and once the concrete has set, you can cut them low with a saw from behind.*

16 An easy way to set levels as you build your steps is to cut a piece of wood to the size of your risers and use it to establish subsequent riser heights. Always check and recheck measurements and use a level, referring to the marks on your measuring post.

15 Dig out the ground behind your first step. Then use a tape measure to establish the position of the next crosstie according to the size of the tread you're using.

And repeat...

Once you've put in the first two steps, it's just about repeating the process: measuring, putting in risers and supports, and continuing like that until you reach the top of the slope.

17 Use your block of timber to check the heights of your risers as you build the remaining steps. Place it on top of the riser below and use a level to set the height for the next step up.

18 Once you've built all the steps and fixed them in place with concrete, predrill two holes per support post and fix them firmly to the sleeper with 4in (100mm) exterior wood screws. Push earth over the concrete to cover it.

19 Once all your steps are in place, the concrete has set, and everything is screwed firmly together, use a handsaw to cut the support posts down 1½–2in (40–50mm) below the top of the riser. They will be concealed by the sod tread.

20 Fill the steps and the surrounding area with soil. Tamp and rake it into place. The soil can be level with the treads as it will settle in time when you lay your sod.

21 Lay sod on each tread and cut it to size with a sharp knife, using the back of the crosstie as a guide. The horizontal plane of the crosstie together with the sod form the tread.

22 Use the back of a flat rake to tamp the sod treads firmly into place. Keep the treads watered in the following weeks so that the grass gets well established. Alternatively, you could sow grass seed.

PICKING PATHS

These picking paths were inspired by the wildlife runs that you find in nature. Once you start looking, you can spot lots of these narrow pathways that rabbits, foxes, and other animals make as a quick route through meadows, woodland, and crop fields.

YOU WILL NEED

Use the instructions on pp244–45 to calculate quantities for:

- crushed stone (2in/50mm to fines)
- mortar mix (1:6 cement to sharp sand)
- bricks for use on the ground
- kiln-dried sand

Tools

- tape measure
- line and pins
- carpenter's square
- lawn edger or flat spade
- wooden pegs
- level
- wheelbarrow
- shovel
- rake
- compacting plate or block of wood
- brick trowel
- masonry saw or angle grinder with diamond disc
- metal rule and sharp-edged tool
- brick cutter or masonry chisel
- gloves
- dust mask
- safety goggles or glasses
- sheet of plastic or cardboard
- soft brush or broom

Picking paths are particularly handy if you've got a big vegetable plot, deep beds with a hedge at the back, or large flowerbeds, and you want a little path to run through them so you can harvest the crops, cut a hedge, water plants, or weed the plot. I've also used them in show gardens as maintenance paths, as they make an attractive detail and are also really handy for accessing a border without treading mud all over the place.

CONSIDERATIONS

- These are made of brick, but you could also make them with strips of stone or wooden edges with gravel in the middle, or simply put bark on top of crushed stone. Alternatively, instead of stone, you could put a layer of concrete in.

- I don't tend to wet point these paths, but you could if you wish. The advantage of wet pointing is that weeds don't grow in the joints, but the disadvantage is that the path won't moss up and look natural so quickly.

- The path uses three bricks in "running bond" pattern, so it ends up being about 12in (300mm) wide. You could, of course, use the same principle and work to your own dimensions.

EXPERT INSIGHT

- These paths are a simple, practical way of getting from A to B, and, because they are narrow, are a great way of getting people to slow down, engage with the space, and appreciate the plants.

- You can use them to introduce different angles from one part of the garden to another and link various spaces. I'll often use them in a woodland setting or through a border to connect one path to another or to a lawn.

- When positioning them, think about how they can help make harvesting crops, picking flowers, or general maintenance easier.

CROSS-SECTION OF THE PATH BASE

It's important to lay a good base for your path. After digging out the site to the correct depth, put down a layer of crushed stone and tamp it down firmly. Then put your mortar bed on top of that to create a firm base for your bricks.

OVERHEAD

Brick edges —

Mortar bed —

Crushed stone —

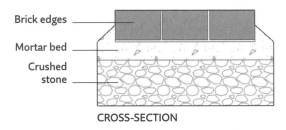

CROSS-SECTION

There are all sorts of creative ways you can use these picking paths in the garden, either to connect different parts or to make watering or maintenance easier.

1 Decide where you want to put your pathway and measure and mark out the sides by setting up your lines to one side. The fixing posts should sit beyond the area you're going to dig out. The path is three bricks across, making it 12in (300mm) wide. Check any right angles are 90 degrees by using a carpenter's square or the 3-4-5 triangle method (see p240).

3 Next, prepare the site by digging out to a depth of around 8in (200mm) in order to accommodate the crushed stone (4in/100mm), mortar bed (1½in/40mm), and bricks (2½in/60mm).

2 Use a lawn edger or flat spade to mark out the edge of your path.

Room to maneuver

It's good to dig a little outside of the lines you've put in to mark the edge of the path. That way you get a bit more space to work in.

5 Use a wheelbarrow to tip your crushed stone into the base of the pathway to a depth of around 4in (100mm), using the line and pins and the pegs you knocked in earlier as a guide.

6 Use a shovel and then a rake to even out the crushed stone around the base of the pathway.

4 Once you've dug to the right depth, firm the base by treading all over it. Knock in wooden pegs to set the crushed stone level at about 4in (100mm) deep. Repeat with more pegs across the base, using a level to make sure they're level. On a path this small you don't need to worry about falls to take the water away, but if I were using the same technique for a larger path, I would allow a fall of ¾in in 39in (20mm in 1,000mm).

Keep things level

After laying the first row of bricks, you might want to move your line and pins over to help guide you when putting in the middle and third rows.

8 You can also use a compacting plate or a block of wood to help ram down the stone if need be.

7 Once the crushed stone is in place, firm it in, working roughly to the wooden pegs, by treading across the site.

9 Make sure all the lines are tight and in the correct place. Mix the sharp sand and cement to make the mortar, and lay a mortar bed along your line. The bed needs to be about ½in (10mm) proud of where the bricks will sit so that you can tamp them down to reach their final level.

10 Lay your first line of bricks lengthways on their widest side. Use the trowel handle to tamp the bricks into place, using the line and a level as a guide. Make sure they're butted up tightly together to minimize the gaps between them.

11 You will achieve a neat finish by using half bricks at the start and end of the middle row to create a staggered pattern. Cut bricks in half with an angle grinder or masonry saw, or score across them with a sharp-edged tool and cut with a brick cutter or chisel.

12 Put in the middle row, starting with your half brick. Use a level to make sure this second run of bricks is level with those in the first row. Check your original line and pins are still level.

Cutting bricks

When cutting bricks, make sure you do it safely, securing the brick firmly in position, and wear protective clothing such as gloves, dust mask, and safety goggles or glasses.

Weather wise

If you're laying your path during wet weather, it's good to have sheeting on hand to cover the works. Also, never lay bricks on a mortar bed in freezing weather.

13 Once you've finished your middle row, ending with a half brick, put in your third row, beginning and ending with a full brick, as with the first row. When the rows are finished and tamped into place, check the levels again. Then use your trowel to put in a haunch (a wedge-shaped mortar support) along both sides.

14 Cover the path with a sheet of plastic or cardboard and leave the mortar to set for a couple of days. Finally, brush in kiln-dried sand to fill all the joints. This will allow the path to moss up over time.

15 I've positioned my picking path through a border to make watering and maintenance easier.

RAISED BEDS

Raised beds are a particularly good way of growing
fruit and vegetables in small gardens as they keep
things neat and tidy. They are also a great way of
growing if you have poor soil as it's easier to improve
what you've got in a contained space, and you can
also adjust it to meet the needs of particular plants.

YOU WILL NEED

- 2 x 2in (50 x 50mm) pressure-treated timber posts, cut to 39in (1,000mm) lengths, 6in (150mm) of which goes in the ground
- 8¼ x 1½in (220 x 40mm) scaffold boards cut to 39in (1,000mm) lengths, with metal end coverings intact
- concrete (1:8 cement to ballast) (optional)
- 3in (80mm) screws
- topsoil and soil improver
- 8 screw-in metal eyes
- string or thin wire

Tools

- tape measure
- builder's square
- spray line or line and pins
- workbench
- level
- pencil
- carpenter's square
- hand saw
- spade
- shovel
- cordless drill/screwdriver
- ¼in (5 or 6mm) drill bit
- hand sledge
- mixing tray or concrete mixer (optional)

Raised beds are straightforward to build—it just involves measuring and cutting the uprights and side boards, then screwing everything together firmly and making sure it's all square and level. If you're using pressure-treated wood, you may want to line the inside of your bed with black plastic sheeting. You can leave the tops of the posts square, as I've done here, or round them off or add finials. This one measures 39 x 39in (1 x 1m) and is 34in (850mm) tall, with 6in (150mm) in the ground.

CONSIDERATIONS

- You can vary the length and width of the beds to suit your site—just think about access from both sides. Ideally, they should be no wider than 4ft (1.2m), so that you can easily reach across them.

- I've made the corner posts tall enough to support mesh or fleece to protect plants, but you can set the height to whatever suits. Taller posts prevent you accidentally dragging the hose over your plants. They are also useful if you grow a crop that needs some support.

- If your soil is poor, dig it out to about 8in (200mm) deep. The boards are also 8in (200mm) deep, so you can put in 16in (400mm) of a mix of 70:30 topsoil to soil improver, such as well-rotted compost or manure.

EXPERT INSIGHT

- The height of the side panels depends on how much you want to spend on materials, the condition of the soil beneath, and how much bending over you're able to do. Remember, the bed is open underneath so roots can access the soil. Around 1ft (300mm) high is suitable.

- Be aware that some scaffold boards are pressure-treated, and others are not. I tend to use treated wood as it lasts longer, but you could opt for untreated hardwood from a sawmill. It will last a decent amount of time but it is relatively costly.

1 Clear the site for your bed, making sure it's free of perennial weeds and the soil is free-draining with no obstructions. Use a tape measure and builder's square (or 3-4-5 triangle method, see p240) to measure the site, and then mark it out with spray line or line and pins.

3 Measure, mark, and cut your four upright corner posts to 39in (1,000mm) tall.

2 Use a carpenter's square, tape measure, and pencil to mark, and then cut the scaffold boards to length. Make sure you have a metal end and a cut end for each side and that the edges are straight.

Scaffold boards

The metal band at each end of a scaffold board helps stop the wood from splitting and makes a nice detail. It's worth keeping any left-over boards as they can come in useful around the garden.

4 Dig out the area, making sure the corners are about 6in (150mm) deep for the uprights. This raised bed measures about 39 x 39in (1 x 1m). If you go any longer than 6ft (1.8m), you'll need to put in extra supports along the middle.

6 Now fix the sawn-edge scaffold boards to the four uprights. Use the 6in (150mm) mark as your guide and make sure the edges are flush. Drill holes at the top and bottom about 1¼in (30mm) in from the edges. Use screws to attach the boards to the uprights.

5 On the end of each of these posts, make a pencil mark 6in (150mm) from the bottom. This is the part that will go below ground. The scaffold boards will sit just above this pencil line.

7 I tend to work on the ground to fix the four sawn ends to the four posts before setting them upright and fixing the ends that have the metal edge. Drill two holes 1¼in (30mm) in from the top and bottom edge of the boards and screw in place. Check angles are square and repeat for the other three corners. Make sure the metal edge covers the sawn edge of the adjoining board.

66 *Having tall corner posts means you can't accidentally drag the hose across the beds, and they're also handy for pulling yourself up after weeding.* 99

9 Fill the raised bed with about 70 percent topsoil and 30 percent organic matter. You can buy in topsoil, but if your garden soil is good enough, use that. As for organic matter, I tend to use well-rotted horse manure, but there are lots of options.

8 Once the four sides are fixed, drop the raised bed into position, making sure the uprights sit 6in (150mm) below ground level. Tap them into place with a hand sledge, then use a level to check that everything is straight and level and adjust if necessary. Firm in the posts by ramming soil around them but take care not to damage the wood. If you want to concrete them in, you should do so at this point.

10 On the two outside edges of the uprights, mark a line 2in (50mm) from the top across each one. Drill a hole ¼in (5mm) deep in the center of each line —eight in total—for the screw eyes.

11 Fix a screw eye in each hole, so there is one on both outside edges of all four uprights.

12 Thread thin wire or string through the eyes and tie off. You can use the wires or string to support netting or fleece thrown over the top of the bed to protect plants from pests or frosts.

Finishing

The timber used here can be left to weather naturally, but you could also try charring it to blacken and preserve the wood or use wood stain.

13 One of the great benefits of raised beds is that they warm up more quickly than open ground, which is really useful for spring plantings. Once you've made one, you'll want to make more.

COMPOST BIN

If you have the space, making your own compost saves money, is good for your plants as it can be used as a mulch or soil improver, and, importantly, it means you can do your bit for the environment by recycling green waste from the kitchen and garden. It's also very satisfying having new life being created from decay.

YOU WILL NEED

- 4 pressure-treated 4 x 4in (100 x 100mm) timber posts about 51in (1,300mm) long
- 6 lengths of pressure-treated roofing batten 39 x 1½ x 1in (1,000 x 38 x 25mm)
- 4 lengths of pressure-treated roofing batten 38 x 1⅜ x 1in (975 x 35 x 25mm)
- 27 pressure-treated timber side boards 39 x 4 x 1in (1,000 x 100 x 25mm)
- 9 pressure-treated timber front boards 38 x 4 x 1in (975 x 100 x 25mm)
- about 8 buckets of concrete (1:8 cement to ballast) or Postcrete
- 2in (50mm) wood screws
- 1½in (40mm) wood screws

Tools

- workbench, vice, or clamps
- tape measure and pencil
- saw
- cordless drill/screwdriver
- line and pins or spray line
- set square
- spade or posthole digger and shovel
- level
- ⅛in (4mm) drill bits
- mixing tray or concrete mixer
- timber roofing battens for props

Ideally, you'd build two or three bins in a row, but if you've only got space for one, then this design is perfect. It's important to build on open ground, not only for drainage but also as it allows the microorganisms from the soil to get into the compost. It is a simple construction made from pressure-treated wood and is about 39in (1m) high, wide, and deep. Three of the sides are fixed, and I've attached battens to the posts at the front to form slots, so that the boards can be dropped in or removed easily. It's a great little project because the skills used can be applied to lots of things, such as building raised beds, archways, or fencing.

CONSIDERATIONS

- Although the perfect place for a compost bin might be right outside the back door, it's not the most beautiful thing. It's better to tuck it away somewhere discreet, such as behind the shed or garage, where it's easily accessible with a wheelbarrow. Ideally, put it in a place where the conditions are fairly constant, such as semi-shade, but avoid anywhere really dark and dry as it will slow down the process of decomposition.

- Compost can take up to two years to reach its best, although once you've mastered the technique, it can be usable after around 12 months. You know it's ready when it becomes a lovely dark brown color and has a crumbly texture.

EXPERT INSIGHT

- You need a roughly 50:50 balance of green—fruit and vegetable waste and other scraps—and brown—dry leaves, cardboard, and small twigs. I adjust the mix when I turn the compost over. Make sure material is in small pieces as this speeds up the process of decomposition.

- Keep the top covered with a sheet of cardboard to retain heat and turn your compost regularly, ideally once a month. This allows air to get in, reduces compaction, and allows you to check the balance of the mix.

1 Prepare all the timber first, ideally on a workbench, so that you can then put it together like a kit. First, cut the four timber posts to 51in (1,300mm) long.

3 Attach the roofing battens along the length of the uprights to provide fixing points for the boards on the sides and back. To make sure everything sits neatly, position a board flush to the edge of the post and sit the batten behind it. Then make pencil marks as your guide. Position the batten 6in (150mm) from the top of the post. Leave the bottom 10in (250mm) of the post free of batten as this part will be dropped into the ground.

2 Cut six roofing battens to 39in (1,000mm) long. These are to support the boards on three sides. Also cut four roofing battens to 38in (975mm) long. These are to create a slot at the front of the bin so that the boards can be removed easily.

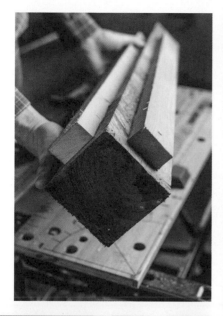

4 You'll need two battens on the back two posts, and one on each of the front two posts. Drill three or four holes along the length of the battens and fix them in place using 2in (50mm) screws.

6 Now you need to create a slot for the front boards so that you have a removable section between the front posts. The front boards will all be 1in (25mm) shorter than the side and back boards so you can slide them in and out easily. Lay your post on your workbench and position one batten flush with the outer edge of the post and the second one 1¼in (30mm) behind it. Place a board between the two to check it will slide easily. Drill three or four holes along the length of each batten and fix in place using 2in (50mm) screws.

5 Cut 27 side and back boards 39in (1,000mm) wide. Then cut nine boards for the removable front panels to a width of 38in (975mm).

Square corners

Use the 3-4-5 triangle method outlined on p240 to make sure the angles are accurate at the corners.

8 Mix your concrete or use Postcrete. Put your first post in one of the holes and use a level to make sure it sits firmly at the base of the hole and is vertical and level. Remember you'll need to fix the sides in between each of the posts so double-check your measurements. Hold the post in place while you fill the hole with concrete, which should be just shy of ground level. Check again that it is vertical and square with your lines. Then use a wooden batten to tamp down the concrete and knock out any air pockets.

7 Make sure the site is roughly level and clear of perennial weeds. Accurately measure and mark out the 39in (1,000mm) square for the base using line and pins, setting the pins about 14in (350mm) wider than the site to allow space to work. Then dig a hole in each corner for the uprights. They should be 12in (300mm) deep and around 12 x 12in (300 x 300mm) square.

9 When your first post is vertical, fix a temporary prop while the concrete sets. Do this by tamping two battens into the ground and screwing them in at an angle, while making sure the post is held vertical. Remember that the top of this post will set the levels for the other posts.

11 Next, predrill two holes, top and bottom, on both ends of the boards for the back and sides. Working from the top down, fix the side and back boards in place with 1½in (40mm) screws. Use your level to check the boards are level, as timber is often not perfectly straight. Make any necessary adjustments.

10 Repeat the process for the three other uprights, making sure they are 39in (1,000mm) apart, are square with each other, and level across the top. Temporarily fix support battens to the sides of the uprights, checking everything is straight as you go. It's worth taking your time with this as it's your last chance to double-check all your levels and measurements. Allow the concrete to set for about 24 hours before removing the support battens.

> *Try to choose a site for the compost bin that's easily accessible with a wheelbarrow, and that's not too dark or dry, as it can slow down the process of decomposition.*

12 Finally, drop the 38in (975mm) boards in the slot between the battens at the front. As these boards are 1in (25mm) shorter than the side and back boards, they should slot into place easily.

BRICK CIRCLE

Circles used throughout a garden can be quite dynamic and, if repeated, can provide rhythm. This brick circle is a great destination point. It could be enlarged and used as a terrace set in lawn or in paving. It works well here in the gravel garden, providing a great focal pull.

YOU WILL NEED

Use the instructions on pp244–45 to calculate quantities for:

- crushed stone
- mortar mix (1:6 cement to sharp sand)
- paving bricks for use on the ground
- pointing mortar mix (1:6 cement to light-colored soft sand)
- sharp sand for bedding in
- kiln-dried sand for filling joints

Tools

- tape measure
- line and pins or spray line
- shovel or spade
- hand sledge
- wooden pegs
- pencil
- rake
- hand rammer or compacting plate
- level
- brick trowel and pointing trowel
- rubber mallet
- hose
- soft brush and broom
- wooden plank for screeding board, 4 x ¾in (100 x 20mm)
- hand saw
- masonry saw or angle grinder with diamond disk
- gloves, dust mask, and safety goggles
- sheet of hardboard
- piece of wood or compacting plate

The first thing is to decide where to position it. I like having brick circles among planting or sitting at the edge of a lawn, half in and half out of plants, so you feel enveloped by the garden. It's only about 6½ft (2m) across, so you wouldn't necessarily use it as your main terrace, but it's great for a seating spot with a good view. If you want to accommodate a table and chairs, build it at least 10ft (3m) across. Circular shapes are easy to build because everything comes from a central point, so you can't really go far wrong.

CONSIDERATIONS

- When you're selecting your materials, think about whether they work with the rest of the garden and the architecture of your house. The terracotta of the bricks I've used here is warm and comforting but doesn't reflect light. You could use lighter colors in lower light areas.

- Be careful not to put this brick pad too close to a back wall or fence as you'll end up with planting spaces that just look awkward. A good rule of thumb is, if in doubt, always make things more generous. Wherever you decide to put it, think about how shapes connect together and feel balanced.

EXPERT INSIGHT

- When you're building any paving that's curved, bear in mind that the smaller the unit of brick you use, the better the curve is going to be and the fewer joints you'll have.

- To work out the area and therefore the quantities of material you'll need, you can revisit some school geometry and use the formula Pi x radius squared (πr^2). A simpler way of doing this is to draw a square over your circle, work out the diameter, and multiply by two. This way you can guarantee you will have sufficient materials for the circle.

CROSS-SECTION OF THE BRICK CIRCLE

The brick circle is laid on a bed of crushed stone and sharp sand, and the brick edging is laid on a bed of mortar on top of the crushed stone. Depending on the size of your circle, you may need a slope for rainwater runoff—refer to the instructions on setting levels on p241.

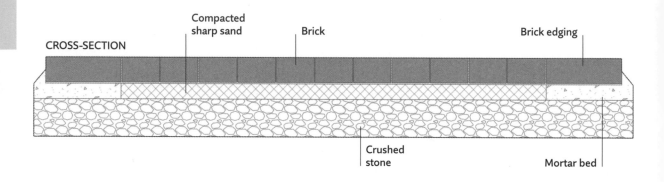

CROSS-SECTION

Compacted sharp sand

Brick

Brick edging

Crushed stone

Mortar bed

BRICK PATTERNS

Once you've decided on a pattern for your bricks, do a practice run by laying some of the bricks in place to check you like the effect and that it will work in your space.

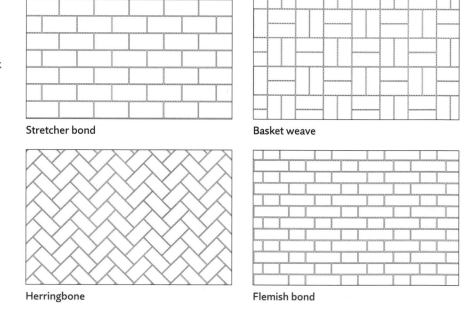

Stretcher bond

Basket weave

Herringbone

Flemish bond

2 Dig out the area to a depth of just over 8in (200mm) to allow for the various layers (see step 4). It's a good idea to dig out the area wider than the 6½ft (2m) final size to have more room to maneuver.

3 Next, make sure there are no soft spots by treading down the base so it's firm and even across the entire site. If treading down is not effective enough, you can use a 4 x ½in (100 x 10mm) post to ram the area.

1 Work out the location and size of your circle and place a pin with a line attached at the center. Use spray line to draw the circle. This one is 6½ft (2m) in diameter but you may want to go up to 10ft (3m).

4 Hammer in pegs across the site, marking 4in (100mm) for the crushed stone, 1½in (40mm) for sharp sand, plus the height of your brick, which is normally about 2½in (60mm).

5 Spread crushed stone to a depth of 4in (100mm). Rake it level and use a hand rammer or compacting plate to ensure the surface is firm and even. Check the pegs around the edge are still set correctly.

6 Using your central pin and level, lay your first brick. Take the line out from the central pin, make sure it runs across the middle of the brick, and then pull it tight. Mark the line or tie a knot at 6½ft (2m) on the line; this will give you your fixed edge to work to. Mix up your mortar and lay down a bed of mortar about 1½in (40mm) thick. Set in your first brick on its side, using the knot as a guide and the level to make sure it's level.

7 Tamp the brick into the mortar with a rubber mallet. Continue putting bricks in around the perimeter, checking the position with the line. If you can, "butter" one side of the bricks as you go. This will help hold the bricks together. If this proves difficult, you can point the bricks at the end by working mortar mix into the joints.

9 Once the brick edging is completed, mix up a cement and soft sand mix (1:6) to wet point between the joints, using a piece of hose to smooth it down. Work the pointing down if you didn't "butter" your bricks. If you pointed the joints when you laid your bricks, it makes sense to work back the mortar every few hours so it doesn't dry out too much. If you find this difficult to do, you can work in your mortar and point once you have finished laying all your bricks in step 18.

8 Continue laying your brick edging, making sure the joints are snug. Keep checking heights and referencing the levels marked on your pegs as you work, and use your line pulled tight through the center of your laid bricks. Placing the bricks on their edge allows you to create a circle with tighter joints between the bricks.

Weather conditions

The ideal conditions are a dry, overcast day. If it's frosty, cover the area overnight to protect the mortar. If it's hot, you don't want things drying out too quickly, so you may need to soak the brick as you work.

10 Brush off any excess mortar. Use a trowel to detail the edge and add a haunch (a ridge of mortar) to the outside of the bricks, keeping it below the top of the bricks so it doesn't show. Once the edging is finished, cover it with a sheet of thick plastic or cardboard to protect it and leave it to set for 24–48 hours.

12 Knock the central peg ½in (10mm) higher than your brick edge, using a level to check it's straight. You'll use the central peg as a pivot for the screed board, moving it in a circle like the arm of a clock, to get the final levels of sand correct. Being slightly higher in the middle means that water will run off and you won't get puddling in the center of the circle.

11 Then cut a piece of timber just over half the radius of the circle to make a screed board. Cut an L-shaped section at both ends. The vertical measurement of the L should be at least ¼in (5mm) less than the depth of your brick, so that after screeding, your brick will sit ¼in (5mm) above the outside edge. Once you have cut your board, it's good to check you will have a good 1½in (40mm) of sharp sand under your bricks.

Brick patterns

There's an almost endless choice for brick-laying patterns, including herringbone, basketweave, and running bricks.

14 When the sand is in place, make sure it's level by pulling your screed board around the radius, using the central peg as a pivot and working backward. As you work, you may need to fill in any holes, so keep a trowel handy to work in small amounts of sand if need be.

13 Next, put a bed of sharp sand on top of the crushed stone and roughly work it into place, levelling it with a rake or shovel. Make sure your sand is a little deeper than your final level to give you extra sand to fill voids as you screed. Eventually, the sand should be at a depth of at least 1½in (40mm), but initially, when you set brick on top, you want them to sit slightly proud. They will settle into place when you tamp them in step 19.

15 Set up a line across the area over the central point to make sure the bricks start out square. This will give you a line to work to. Start setting your bricks on the sand and butt the bricks up to each other closely. I'm doing a herringbone pattern but there are lots of alternatives. You could also simply fill the central space with gravel or paving, whatever your taste. Take your time laying bricks and remember to stay off the screeded sand.

17 Use a masonry saw or angle grinder with diamond disc to cut along the marked line. Hold the brick in place using timber. Repeat the process for the other bricks, tamping them into place in the circle.

16 When you've finished the main area with full-sized bricks, there will be gaps around the edge that you'll need to fill with cut bricks. First, put the brick into position. Then, using the edge of the circle as your guide, draw the shape onto the brick.

This brick-edged gravel circle is slightly wider at about 10ft (3m). Filled with gravel instead of bricks and surrounded by established planting, it's an ideal spot to put a table and chairs.

18 Once all the bricks are in place, pour kiln-dried sand over the top and use a soft broom to sweep it into the gaps between the bricks. The sand will settle in place over the coming weeks and may need topping up at some point.

> *Small brick pads are a great way to create destination points in the garden, particularly if they are surrounded by plants or offer a good view.*

19 To tamp the bricks down into their final position, I use a big block of wood and a rubber mallet and work across the surface. Or you can use a compacting plate, but you will need to cover the area with a sheet of hardboard before you start. The brick paving should now be level with the edging. I leave a little sand on the surface, which will settle into place as the whole thing beds down.

20 The brick pad looks great in a wider gravel area as a place to put a fire pit, container pots, or a small seat.

USEFUL
INFORMATION

MARKING OUT THE SITE AND SETTING LEVELS

An important part of any building project is to mark out your site accurately. Whether you're transferring information from a plan or marking it straight to your site, it can sometimes feel more daunting than it really is.

Note that in this book I've included both imperial and metric measurements. Whatever system you use, make sure you're consistent and use either all imperial or all metric measurements for individual projects. Here, I've used metric units only for some calculations to keep things simple.

MARKING OUT THE SITE

If you need to dig out foundations, the vertical markings on your pegs will determine how deep you need to dig for the various layers, such as cobbles, mortar, paving, gravel, brick, and so on. If you need to incorporate a slope to ensure rainwater runs off, then factor this in too (see Setting a level for rainwater runoff, opposite).

1 Roughly mark out the area using string and pins or spray line. If you're creating a rectilinear shape, use a builder's square or the 3-4-5 triangle method (see below left) to check right angles are an accurate 90°. I generally make the area about 8in (200mm) bigger than I need to give me a bit of additional space to work in.

2 Dig out the area, using either a spade or a mini-digger. I tend to dig down to 6–8in (150–200mm), although this may vary depending on the project and materials I'm working with. Once you're at the right depth, tread down the soil across the marked area so it's firm and even. If your heel sinks into any soft patches, dig down further until you reach a hard surface.

3 Put the first pegs around the perimeter of the site, making sure they are at the same height as you want your finished surface to be. Then put in more pegs across the entire site and use a level to check they are at the correct levels. If you need to allow for a slight slope for rainwater to drain easily, refer to Setting a level for rainwater runoff, opposite.

4 Using the finished surface level as your guide, measure down where you want the levels of the cobbles, mortar, paving, gravel, brick, and so on to sit, and use a pencil to mark this on the pegs. For cobbles, aim for a depth of no less than 4in (100mm). Getting these levels right is really important so keep checking as you go.

3-4-5 TRIANGLE

Making sure that your corners are accurate is one of the biggest challenges when building. One way to create an accurate 90° angle is to use the method known as the 3-4-5 triangle. This simple technique is based on a theory established by Greek mathematician Pythagoras. He worked out that the measurements of a right-angled triangle are in the ratio of 3 on one side, 4 on the other, and 5 on the longest side. If you use line and pins to form a triangle that measures 300mm, 400mm, and 500mm (or multiples of these numbers), say, you can be sure that you have an accurate right angle between the two shorter sides.

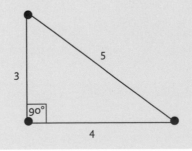

SETTING A LEVEL FOR RAINWATER RUNOFF

A slight slope on your paving or brick surface will allow rainwater to drain easily and prevent puddling or moss building up. In general, paving with a rough or textured surface needs a steeper slope than paving with a smooth finish.

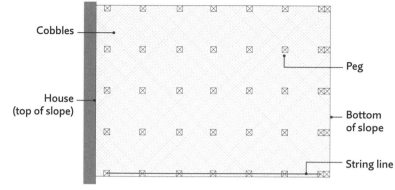

SETTING A SLOPE OVERHEAD VIEW

Cobbles

House (top of slope)

Peg

Bottom of slope

String line

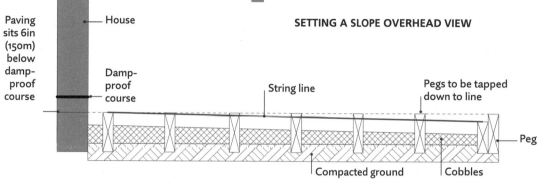

Paving sits 6in (150m) below damp-proof course

House

Damp-proof course

String line

Pegs to be tapped down to line

Peg

Compacted ground

Cobbles

SETTING A SLOPE SIDE VIEW

When you buy materials, you'll often find the recommended gradient written on the packaging, but if not, ask your supplier. The slope, fall, or gradient is expressed as a ratio, and you'll see recommendations of anything between 1:40 to 1:100, which means a fall of 1 unit for every 40 or 100 units of distance. If you live in an area that has high rainfall, it's worth putting in a slightly steeper slope.

After you've excavated the area, set up pegs on the perimeter to the level you want your final surface. If you're building next to a house, the paving must sit 6in (150mm) below the top of the damp-proof course. You should be able to spot a black liner at the base of the house—this is the damp-proof course.

For a patio, the standard fall is 1:80, so for every 80 units of distance, the level should fall by 1 unit. If your patio runs 6½ft (2m) from the house, for instance, you should divide 6½ft (2,000mm) by 80, which is 1in (25mm). Make sure that the top of your slope is next to the house and inclines down away from it, so the patio should be 1in (25mm) lower at the side furthest from the house.

1 Begin by knocking in pegs in straight lines across the site, working away from the house. You need a row where the end of your slope will sit. The pegs should be at the eventual level of your finished paving.

2 Use a pencil to mark the 1in (25mm) fall on the pegs at the bottom of the slope, 6½ft (2,000mm) away from the house. The fall should always be even across the site, so use the longest level to get your levels straight horizontally.

3 Next to these bottom pegs, knock in pegs to the height of the 1in (25mm) pencil mark.

4 Fix a string line at right angles to the house from a peg at the top of the slope to one of the lower pegs at the bottom. Make sure it is tight. Working back to the top of the slope, tap down the other pegs along the line so they are flush with the line.

5 Repeat across the site and use a level to check levels horizontally across the site. Measuring from the top of the pegs, mark the level of the cobbles on the pegs.

If you have more complex level changes in your garden, it's a good idea to hire some equipment, such as a laser-level device with a measuring staff. It's quite straightforward to use, and it will give you accurate readings around your space. Put it on a fixed point in the garden, such as a drain cover, and take your readings from there.

YOU WILL NEED

- tape measure
- line and pins or spray line
- hand sledge
- carpenter's square or use 3-4-5 triangle method
- small and large levels
- spade
- shovel
- mini digger (optional)
- sod cutter (optional)
- wheelbarrow
- wooden pegs/battens for marking levels (have plenty of these)
- rake
- compacting plate, whacker plate, or hand rammer
- iron bar (optional)
- gloves
- safety goggles
- face mask
- first aid kit

CREATING A GOOD BASE

Once you've mastered the core skill of building foundations, you can use it for all sorts of projects. Making a strong, level base is key when putting down paving, gravel, paths, or bricks. It's well worth putting the time in to do this well, as poor groundworks often lead to the paving or walls eventually moving and potentially cracking.

The best way to build a good base is to dig out the looser topsoil until you reach a firm layer of subsoil. You then need to fill the base with a layer of cobbles, which you compact down before adding a layer of sharp sand or possibly a bed of mortar on top. For brick walls, you need to pour in concrete to about 12–24in (300–600mm) deep, depending on what you're building, to create a solid foundation.

PLANNING

If you're working from a plan, once you have transferred your measurements to the ground, check that everything is going to work. If you've made a scale drawing, then use that as your guide. This is often your last opportunity to change things, so check the size of the space works for its intended use and that you're happy with the materials. If you need to factor in a slope to ensure rain drains easily, work that out now (see Setting a level for rainwater runoff on page 241).

When working out the depth you need to dig, remember that for paving—either bricks or slabs—you need to allow for the cobbles, mortar mix, and paving material. For gravel or bark, you need to allow for cobbles and the thickness of the top layer. For a low brick wall, you'll need to pour in a concrete footing. See Calculating Materials on pages 244–45.

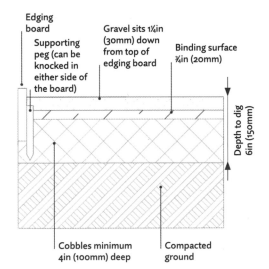

Edging board

Supporting peg (can be knocked in either side of the board)

Gravel sits 1¼in (30mm) down from top of edging board

Binding surface ¾in (20mm)

Depth to dig 6in (150mm)

Cobbles minimum 4in (100mm) deep

Compacted ground

FOR A GRAVEL PATH WITH EDGING BOARDS

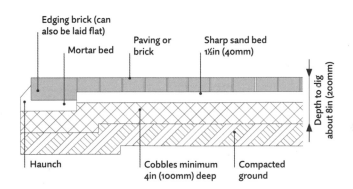

Edging brick (can also be laid flat)

Mortar bed

Paving or brick

Sharp sand bed 1½in (40mm)

Depth to dig about 8in (200mm)

Haunch

Cobbles minimum 4in (100mm) deep

Compacted ground

FOR BRICK PAVING

SOURCING

Calculate the quantities you'll need for: cobbles, sharp sand, or cement (see Calculating Materials on pages 244–45), again using your scale drawing if you have one. I tend to make an allowance of about 10 percent extra for wastage, but this does depend on the material; your supplier should be able to advise. Make sure you have plenty of wooden pegs/battens to mark out levels.

1 Do a final check to make sure all pegs across the site are at the correct level. If this is the first time you're doing this sort of building work, then use lots more pegs to give yourself plenty of reference points.

2 Using the pegs as your guide, tip the cobbles into the area you've dug out, spreading them evenly across the site.

3 Rake the cobbles then tamp them down with a hand rammer or compacting plate so that you end up with a firm, even, and level surface. Work around the pegs, leaving them in place as they indicate the remaining levels. If you have any sunken areas, you should add in more cobbles and tamp them down. The site is now ready for the next stage, whether it's adding a bed of sharp sand, gravel, or a mortar bed.

CALCULATING MATERIALS

Before you start on any project, it's important to have an accurate measure of the quantities of materials you'll need. Make sure you measure the dimensions of the proposed site carefully, and use the advice outlined here to make your calculations.

Note that the calculations in this section are all approximate. Speak to your building materials supplier to get advice on quantities for specific materials, or check all your measurements on an online calculator if you're ordering online.

Metric conversions

If you need to convert mm to m, remember that:

1m = 1,000mm
1 square meter (m²) = 1,000,000mm²

BRICKS FOR A PATHWAY OR BRICK CIRCLE

To calculate how many bricks you'll need for a pathway, terrace, or brick circle, use these simple calculations. Most brick suppliers will also tell you how many bricks you need to cover a square meter.

First, calculate the area of the site, which is width x length, to give you the square meterage (m²). Next, work out the area of 1 brick (width x length).

Then divide the area of the site (the square meterage) by the area of 1 brick and add 10 percent for wastage to give the total number of bricks needed.

Example 1 picking path

The area of a pathway is 0.3m (300mm) x 5m = 1.5m².
The area of 1 brick is 0.215m (215mm) x 0.1025m (102.5mm) = 0.02204m².
The number of bricks needed for 1.5m² is 1.5/0.02204m² = 68, rounded up to 70.
Add 10% (7) for wastage:
70 + 7 = 77 bricks needed in total.

Example 2 brick circle

To work out the quantities of material you'll need for a circular brick pad, you can calculate the area of the circle by using the formula Pi x radius squared (πr^2). However, a simpler way of doing this is to calculate the area of a square that has the same length and width as the diameter of your circle.

The diameter of a circle is 2m. The area of a 2 x 2m square = 4m².
The area of 1 brick is 0.215m (215mm) x 0.1025m (102.5mm) = 0.02204m².
The number of bricks needed for 4m² is 4/0.02204mm² = 181, rounded up to 190.
Add 10% (19) for wastage:
190 + 19 = 209 bricks needed in total.

COBBLES, GRAVEL, AND **SHARP SAND**

Since these materials involve depth as well as area, you need to work in cubic units to calculate volume. Volume = width x length x depth.

Example

The area of a terrace is 3 x 4m = 12m². To calculate the volume of cobbles, gravel, or sand needed to lay it to a depth of 0.1m (100mm), multiply the area by the depth: 12 x 0.1 = 1.2 cubic meters (m³).

MORTAR BED FOR **PAVING**

As a ballpark figure, I allow 8 to 10 bags of cement for 1m², but obviously it depends on what sort of mix you're making (1:10; 1:8, and so on). To calculate the volume of sand and cement you need to make mortar, divide the total volume by the sum of the ratio numbers. Then multiply that figure according to the ratios needed.

Example

The area of a terrace is 3 x 4m = 12m². To lay mortar to a depth of 0.04m (40mm), multiply the area by the depth: 12 x 0.04 = 0.48m³.

You then need to work out the quantities of cement and sand, depending on the ratio of your mortar mix. For a 1:6 mix of cement to sand, divide the total volume by 7 (6 + 1): 0.48/7 = 0.0686m³. So, 1 part cement is 0.0686m³ and 6 parts sand is 6 x 0.0686 = 0.412m³.

CIRCULAR POOL

It is easiest to calculate materials if you use the area of a square that has the same length and width as the diameter of your pool. For liner and protective fleece, I tend to allow 0.5 to 1m on each side as extra.

Add twice the maximum depth to the length of your square plus the extra. Then add twice the maximum depth to the width of your square plus the extra. Multiply the length total by the width total to get the area of liner and protective fleece you need.

Example

The diameter of a pool is 3m and the maximum depth is 1m. Allow for 1m extra either side.
Length is 3 + (2 x 1) + 2 = 7m
Width is 3 + (2 x 1) + 2 = 7m
The area of liner and protective fleece required is 7 x 7m = 49m².

For cobbles, allow about one 20kg bag per 0.4m². Divide the area of your pool by 0.4 for the number of 20kg bags you will need. This is only a rough guide and will, of course, change according to how deep you want your covering.

Example

The diameter of a pool is 3m. The area is 3 x 3m = 9m².
9/0.4 = 22.5, rounded up to 23.
23 x 20kg bags (or 460kg) needed.

CONCRETE FOOTING FOR **POSTS**

For the concrete footing for posts, calculate the volume of each hole (width x length x depth). Then multiply the volume by the number of holes, or posts, you plan to have.

Example

There are 10 posts. Each post hole is 0.3m (300mm) x 0.3m (300mm) x 0.6m (600mm) = 0.054m³.
The amount of concrete needed is 10 x 0.054m³ = 0.54m³.

GLOSSARY

AGGREGATES

Aggregates are loose materials that can be mixed with cement to form concrete or mortar. They include sand, gravel, and ballast. *See also* **Ballast, Concrete, Fines, Mortar**.

AQUATIC PLANTS

Aquatic plants are plants that have adapted to live in water—either fresh water or salt water. Different plants have different requirements regarding the depth of water they will grow in. Marginal aquatic plants grow in shallow water, near the edge of a pond. Deep-water aquatics live in the deeper water, usually near the middle of a pond, while floating plants float freely in the water. Many provide oxygen for the water and habitat for wildlife. *See also* **Oxygenators**.

BALLAST

Ballast is a type of coarse aggregate that is made up of a mixture of sand and gravel, with smaller and larger particles. It is used in combination with cement and water to form concrete. The proportion of cement to ballast can range from 1:6 to 1:12, depending on its use.

BATTEN

Batten is a strip of solid material, usually wood, that is used to provide fixing points or to reinforce a frame. Roofing lath (also called roofing batten) are small sections of treated timber often used as batten.

BINDING SURFACE

This is often used to help bed gravel when cobbles are very hard and do not give. I usually use ballast or hoggin (a self-binding gravel) if I'm laying a gravel path.

CLIMBERS

This term refers to plants that grow upward by being attached to other plants or objects. Some support themselves using tendrils or suckers, while others will need to be tied onto a supporting frame such as a trellis. *See also* **Ramblers**.

COBBLES

This term refers to the solid material used as a filling or to create a solid base for paths and terraces. It is made up of pieces of bricks, crushed quarry waste, rubble, and gravel. It ranges in size from large pieces to dust. You may also hear it called granite or limestone, depending on where you live.

CONCRETE

Concrete is a mixture of water, aggregates, and cement. The mixes change depending on use. It is generally used for footings beneath walls and to hold posts in place, but it can also be used in ornamental ways.

FINES

When aggregate is sieved, the particles measuring usually ⅕in (4.75mm) or smaller are referred to as "fines." This fine aggregate is used to fill the gaps in coarse aggregate and to help to create a firm surface.

FOOTING

A footing is a solid, durable foundation used beneath the base of a brick wall or steps to make sure everything remains stable. Normally, it is formed by digging a trench, filling it with concrete, tamping it down until it's level and there are no air pockets, and then leaving it to set hard for at least 24 hours, but normally longer.

HARD LANDSCAPING

Also called "hardscape," this term refers to the built elements of a garden or landscape. It includes things such as paved areas, driveways, walkways, steps, and walls that are made from hard-wearing materials such as stone, wood, brick, concrete, steel, or gravel.

HAUNCH

A haunch is a small ridge of mortar added to the side of edging bricks or a brick wall to help to hold the bricks in place.

HEDGING

A hedge is a line of closely planted shrubs or trees, evergreen or deciduous, that knit together to form a barrier.

HERBACEOUS PERENNIALS
see **Perennials**

KILN-DRIED SAND
This is a clean, fine, dry sand that is brushed into the joints between brick paving.

MARGINAL PLANTS
see **Aquatic plants**

MORTAR
Mortar is a mix of cement, sand, and water that is used between brickwork, under paving, and for edging bricks.

OXYGENATORS
Oxygenators are submerged aquatic plants that produce oxygen during the day and provide cover for aquatic life. They help to keep the water in a pond clean, clear, and oxygenated.

PAVERS
Generally, these are small rectangular blocks of quarried stone made for paving roads and paths.

PEBBLES
These are naturally rounded stones that are available in a range of colors. They can be used to create a natural-looking border around ponds and other water features.

PERENNIALS
Essentially, these are plants that persist for several growing seasons. Some are short-lived, while others can be long-lived. Herbaceous perennials have nonwoody stems, and they typically die back in winter. Their roots survive below ground, and the plant regrows in spring. Woody perennials, such as trees and shrubs, have stems that survive above ground all year round. Some are classed as evergreen, and others are deciduous, losing their leaves in winter.

RAMBLERS
This term is usually applied only to a group of vigorous climbing roses, known as rambler roses. Some grow up to 20ft (6m) tall, and they need a good supporting frame. They usually flower only once in early summer, unlike climbing roses, which can keep flowering almost all through summer and into fall.

ROOFING LATH see **Batten**

SCALE
When designing a garden, we use scale to translate measurements taken from the actual landscape and reduce them to a workable size that we can fit on a piece of paper. This means we can then easily design the whole space, as well as use the measurements to work out useful information, such as quantities needed. It is usually shown as a ratio, for example, 1:50 or 1:100.

SCREEDING
This term refers to the practice of evening out a layer, such as sand, with a screeding board—usually a length of timber cut to size—which is pulled across the surface to smooth it.

SHARP SAND
This coarse, gritty sand is used for mortar mixes for bedding in paving slabs or brick paving or underneath brick edging.

SOD
Sod is basically grass and the layer of earth held together by roots beneath it. You can buy rolls of sod to lay a lawn.

SOFT LANDSCAPING
Soft landscaping, also known as "softscape," describes the horticultural elements used in garden and landscape design, including trees, shrubs, perennials, grass, and bulbs.

SOFT SAND
Also known as builder's sand or bricklayer's sand, it is used in mortar mixes for walls and steps and for pointing brickwork.

SUBMERSIBLE PLANTS see **Oxygenators**

TAMP
To tamp is to compact, compress, flatten, and knock out any air pockets. Tamping ensures a surface, such as sand or concrete, is even and level.

WET POINTING
This term refers to the practice of filling the joints between bricks or paving stones with wet mortar to ensure the gaps are sealed and neat.

INDEX

SUPPLIERS

Below is a list of my favorite suppliers who have supplied the products used in the projects in this book. You will no doubt have your own favorites, but I'm always keen to learn from others and to add to my list whenever I get a recommendation from a trusted source.

PLANTS

Hardy's Cottage Garden Plants
Priory Lane, Freefolk
Whitchurch RG28 7FA
Tel: 01256 896533
www.hardysplants.co.uk

Surreal Succulents
Tremenheere Sculpture Gardens
Gulval, Longrock,
Penzance TR20 8YL
Tel: 07476 349545
www.surrealsucculents.co.uk

Swines Meadow Farm Nursery
47 Towngate E, Market Deeping
Peterborough PE6 8LQ
Tel: 07432 627766
www.swinesmeadowfarmnursery.co.uk

SEEDS/COMPOST ETC

Marshalls Garden
Alconbury Hill
Huntingdon PE28 4HY
Tel: 01480 774555
www.marshallsgarden.com

Westland Seeds
www.westlandseeds.com

LINSEED PAINT

Brouns & Co.
Highfield, Selby Road
Garforth LS25 2AG
Tel: 01423 500694
www.linseedpaint.com

ANTIQUE POTS AND TABLE

The Old Bakery Antiques
23 Main St, Wymondham
Melton Mowbray LE14 2AG
Tel: 01572 787472
www.antiques-atlas.com

GARDEN POTS

Whichford Pottery
Whichford
Shipston-on-Stour CV36 5PG
Tel: 01608 684416
www.whichfordpottery.com

ADAM'S SHIRTS

Paul Smith
www.paulsmith.com

POND LINER/AQUATIC PRODUCTS

Water Gardening Direct Ltd
Hards Lane, Frognall
Deeping St James
Peterborough PE6 8RL
Tel: 01778 341199
www.watergardeningdirect.com

TURF

Lindum Turf
West Grange, Thorganby
York YO19 6DJ
Tel: 01904 448675
www.turf.co.uk

Rolawn
The Airfield, Seaton Ross
York YO42 4NF
Tel: 01904 757372
www.rolawn.co.uk

TOOLS

Classic Hand Tools Limited
Unit B Hill Farm Business Park
Witnesham IP6 9EW
Tel: 01473 784983
www.classichandtools.com

STONE AND BRICK

Branch Bros Limited
106 Bridge Street, Deeping St James
Peterborough PE6 8EH
Tel: 01778 342255
www.branchbros.co.uk

Stamford Stone Co. Ltd
Swaddywell Quarry, Stamford Road
Peterborough PE6 7EL
Tel: 01780 740970
www.stamfordstone.co.uk

York Handmade Brick
Winchester House, Forest Lane
Alne, York YO61 1TU
Tel: 01347 838881
www.yorkhandmade.co.uk

ACKNOWLEDGMENTS

FROM ADAM

Well, thank you all. This booked turned out to be a whole lot bigger than I thought, and I could not do the things I get to do without having some wonderful people around me. So when it comes to thank yous, where do I start? Maybe with my mates, Dave, aka Boat-maker, Will, David, Luke, Shane, and Dan for lending a hand and a laugh when needed. Abbie-Jade, Polly, Sally, and Cob for the help in the office, Juliet for pulling all of this stuff from my head, and Ruth and Barbara for sorting all the things I'm no good at. Jason, my friend, you capture things beautifully, and, of course, Mrs. Frost, for just being you x.

P.S. Sorry to anyone I have missed.

FROM THE PUBLISHER

DK would like to thank everyone in Adam's team, including Abbie Frost and Polly Hindmarch, for all their behind-the-scenes help creating the book. Thanks also to Jason Ingram for photography, Adam Brackenbury for retouching work, Simon Maughan for reviewing the text, John Tullock for US consulting, Eloise Grohs for design assistance, Jane Simmonds for proofreading, and Vanessa Bird for indexing.

DK | Penguin Random House

Project Editor Juliet Roberts
US Editor Sharon Lucas
Senior Editor Dawn Titmus
Project Designer Vanessa Hamilton
Senior Designer Barbara Zuniga
Jacket Coordinator Jasmin Lennie
Jacket Designer Amy Cox
DTP and Design Coordinator Heather Blagden
Production Editor David Almond
Senior Production Controller Luca Bazzoli
Editorial Manager Ruth O'Rourke
Design Manager Marianne Markham
Consultant Gardening Publisher Chris Young
Art Director Maxine Pedliham
Publishing Director Katie Cowan

Photographer Jason Ingram

Royal Horticultural Society
Editor Simon Maughan
Publisher Rae Spencer-Jones
Head of Editorial Tom Howard

First American Edition, 2022
Published in the United States by DK Publishing
1450 Broadway, Suite 801, New York, NY 10018

Copyright © 2022 Dorling Kindersley Limited
A Division of Penguin Random House LLC
Text copyright © Adam Frost 2022
22 23 24 25 26   10 9 8 7 6 5 4 3 2 1
001–322493–Mar/2022

A catalog record for this book is available
from the Library of Congress.
ISBN 978-0-7440-4816-2

Printed and bound in China

www.dk.com

For the curious

This book was made with Forest
Stewardship Council™ certified paper –
one small step in DK's commitment to a
sustainable future. For more information
go to www.dk.com/our-green-pledge

"This one's for Uncle Greg. Thank you for your time, pal. x"

ABOUT THE AUTHOR

Adam Frost is an award-winning garden designer, with seven Royal Horticultural Society (RHS) Chelsea Flower Show Gold Medals to his name. He is a presenter on BBC *Gardeners' World* and BBC coverage of the RHS Flower Shows.

He started his gardening career as a 16-year-old gardening apprentice with the North Devon Parks Department. He then moved back to London to train as a landscape gardener, before getting a job with the late, great gardener Geoff Hamilton. During his time with Geoff, he trained as a landscape designer and then established his own garden landscape business in 1996, which has taken him around the world designing gardens.

In 2017 he set up The Adam Frost Garden School at his home in Lincolnshire, where he hosts informal workshops for gardening enthusiasts.

Adam is also an RHS Ambassador, carrying out a mission close to his heart: to encourage our next generation of gardeners and to raise the profile of careers in horticulture.

Adam and his wife, Sulina, live in a picturesque Lincolnshire village with their four children, two horses, two dogs, three sheep, and Ash (his cat sidekick). His five-acre garden is regularly featured on *Gardeners' World* as a work in progress.